Ansgar Geiger

Strategic Power Plant Investment Planning under Fuel and Carbon Price Uncertainty

Strategic Power Plant Investment Planning under Fuel and Carbon Price Uncertainty

by
Ansgar Geiger

Dissertation, Karlsruher Institut für Technologie
Fakultät für Wirtschaftswissenschaften
Tag der mündlichen Prüfung: 27.04.2010
Referent: Prof. Dr. Stefan Nickel
Korreferent: Prof. Dr. Wolf Fichtner

Impressum

Karlsruher Institut für Technologie (KIT)
KIT Scientific Publishing
Straße am Forum 2
D-76131 Karlsruhe
www.ksp.kit.edu

KIT – Universität des Landes Baden-Württemberg und nationales
Forschungszentrum in der Helmholtz-Gemeinschaft

KIT Scientific Publishing 2011
Print on Demand

ISBN 978-3-86644-633-5

Strategic Power Plant Investment Planning under Fuel and Carbon Price Uncertainty

Zur Erlangung des akademischen Grades eines
Doktors der Wirtschaftswissenschaften
(Dr. rer. pol.)
von der Fakultät für
Wirtschaftswissenschaften
des Karlsruher Institut für Technologie (KIT)
genehmigte

DISSERTATION

von

Dipl. Wi.-Ing. Ansgar Geiger

Tag der mündlichen Prüfung: 27.04.2010
Referent: Prof. Dr. Stefan Nickel
Korreferent: Prof. Dr. Wolf Fichtner
2011 Karlsruhe

Acknowledgments

This work was done with the financial support of the Department Optimization of the Fraunhofer Institute for Industrial Mathematics (ITWM).

First, I would like to thank Prof. Dr. Stefan Nickel for taking over the supervision of this thesis. While he supported me, he gave me the necessary freedom to pursue my own ideas.

Furthermore, I thank my former project partners, especially Rüdiger Barth from the Institute of Energy Economics and the Rational Use of Energy of the University Stuttgart, for the interesting discussions we had.

I am grateful to Michael Monz and Uwe Nowak for their advice concerning software design and the efficient implementation of algorithms. In addition, I thank Richard Welke for his LaTeX tips. Furthermore, I would like to thank Michael Monz for the ideas he gave me for my work.

I am especially indebted to Stefan Reicherz, who took the time to read the whole thesis and who gave me many valuable remarks. Furthermore, I thank Martin Berger, Helmuth Elsner, Hendrik Ewe, Matthias Feil, Ingmar Geiger, Tabea Grebe, Volker Maag, Uwe Nowak, Heiko Peters, Manuel Scharmacher, Jörg Steeg, and Richard Welke for proof-reading parts of the thesis and giving me their advice.

Finally, I thank all who supported me through the times of this thesis, especially my family.

Kaiserslautern, December 2009

Contents

i

List of Figures

List of Tables

Preface

Since the invention of the electric light bulb by Thomas Edison in 1879, electricity has steadily gained in importance. Nowadays, a life without electricity is hardly imaginable. Almost all modern technical equipment like machines or computers relies on electricity. Worldwide electricity generation increased from 11,865 TWh in 1990 to 20,202 TWh in 2008 (BP [BP09]). The revenues generated by electric utilities emphasize the importance of this sector: in 2008, these revenues were over 61 billion € in Germany (Destatis [Des09b]). Associated with these high revenues are high investment costs: a new combined cycle gas-turbine (CCGT) plant costs about 500 million €, a new coal plant about 1,000 million €. Hence, investment decisions must be taken with care.

Besides its importance for the whole economy, another important characteristic of the electrical power industry is its exposure to different kinds of uncertainty. In the past, the two main types of uncertainty have been the uncertainty about input costs and the uncertainty about future legislation.

A large part of electricity is produced by fossil fuels—coal, lignite, natural gas, and, to a lower extent, oil. Fuel prices exhibit significant volatility, especially those for oil and gas, as it can be seen in Figure 1. Even though these uncertainties have always existed, they were not very problematic in the past, since the electrical

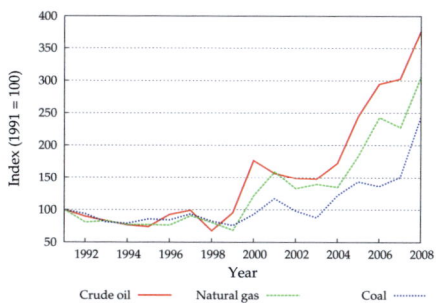

Figure 1: Average yearly import prices in Germany. Source: BMWi [BMW09].

power companies could pass the risks to the consumer. This was possible due to the special structure of the electricity market: the electricity market is a well known example of a natural monopoly.

To be able to deliver electricity, a transmission and distribution network is required. These networks were built by integrated electricity companies. Such integrated electricity companies cover the whole value added chain of electricity: generation, transmission, distribution and sales. The dependence on a network to deliver electricity forms high barriers to entry for potential competitors. The result have been regional monopolies for electricity companies.

These natural monopolies existed for a long time and were not considered as something bad. To limit the power of the monopolists, the electricity market was strongly regulated. Concession contracts were concluded between municipalities and electricity producers. The main objective of these contracts was to guarantee a high level of security of supply. In addition, these contracts should prevent competing investments in transmission networks, which seem to be undesirable from an economic point of view (Wiedemann [Wie08]). Therefore, municipalities were allowed to grant the exclusive right for electricity supply to an electricity company within a fixed area. In exchange, the electricity company had to ensure a certain level of security of supply. While monopolies are generally prohibited, they have been allowed in the electricity sector. In Germany, this was regulated by §103 of the Act Against Restraints of Competition.[1] As a result, consumers had no choice concerning their electricity supplier, and consequently, the electricity industry was characterized by a low level of competition.

With the beginning of the 1980s, the attitude towards the monopolies in the electricity industry slowly changed. While monopolies in the case of electricity transmission and distribution still seemed to be justified, those in the case of generation and retail were questioned. Chile was the first country to introduce privatization and other market-oriented reforms in the early 1980s. Other Latin-American countries followed the example (Sol [Sol02]).

In Europe, the United Kingdom was the first country that privatized its electricity supply industry in 1990 and created competition in the retail sector. The next European country that introduced an electricity market was Norway in 1991 (OECD and IEA [OI01]). In 1996, the Council of the European Union adopted the directive 96/92/EC concerning common rules for the internal market in electricity. This directive imposed an opening of the electricity market. In Germany, the directive

---

[1]Gesetz gegen Wettbewerbsbeschränkungen (GWB)

was implemented with the amendment to the Energy Industry Act[2] in 1998. Vertically integrated electricity companies were obliged to keep separate accounts for their generation, transmission and distribution activities. Access to the transmission and distribution system must be granted applying a non-discriminatory tariff.

It is often questioned whether the liberalization of the electricity market imposed the desired level of competition (Newbery [New02], Müsgens [Müs04], Hirschhausen et al. [Hh+07]). Even if the intensity of competition can still be increased, the deregulation certainly led to a higher level of competition. With increased competition, good planning tools gained importance. While in the days of a regulated market, the complete fuel price risk could be passed to the consumer via increased electricity prices, this is no longer the case. Customers may change their electricity supplier, if they are not satisfied with its prices. This happened to Vattenfall in Germany after they announced a price increase in July 2007 (Manager Magazin [Man07]). Besides the fuel price risk, generation companies are now additionally exposed to two other forms of risk: volume risk, referring to the amount of electricity sold, and electricity price risk. Nowadays, it is very important for the electricity generation companies to minimize the impact of the uncertainties they are facing.

The importance of good planning tools, combined with the challenges caused by the uncertainties and the financial importance of the electricity sector, attracted a lot of research. During the time of a regulated market, the focus of the research was on the optimization of electricity generation. For this purpose, the electricity market with its units, the electricity demand, and, in some cases, the transmission network, was modeled. The objective was a cost minimal electricity production subject to demand satisfaction and—depending on the level of detail modeled— several technical constraints. Various optimization techniques varying from linear programming (LP), mixed-integer linear programming (MILP) to different metaheuristics like genetic algorithms or simulated annealing have been applied to this problem. Extensive studies of these methods have been carried out by Sheble and Fahd [SF94] and Padhy [Pad04].

The purpose of these models was twofold: on the one hand, they could be used to determine the cost-minimal unit commitment for the generation company. On the other hand, they could be used from an economic point of view. As such, these

---

[2]Energiewirtschaftsgesetz(EnWG)

fundamental electricity market models could be applied by the regulator to determine an adequate level of electricity prices, which correspond in a competitive market to the marginal costs of the most expensive unit producing electricity.

Due to the deregulation of the electricity market, another problem gained in importance: the power generation expansion problem. In a regulated market, investments in new power plants were carried out such that the certain level of security of supply could be guaranteed. The focus of the investment decision lied on the choice of the technology. This decision was taken based on the levelized costs of electricity production, which denote the expected average costs per MWh. In a deregulated market, this method is no longer pertinent, as it does not consider uncertainties related to future costs and profits in an adequate way.

A popular method for investments under uncertainty is the real options approach (see, e.g., Dixit and Pindyck [DP94]). It is based on the idea that under uncertainty, the option to postpone an investment has a value, which must be considered for the investment decision. This option value is related to new information that becomes available in the future, and may influence the investment decision.

With the deregulation of the electricity market, the application of the real options theory to power plant valuation became popular (e.g., Deng [Den99; Den05], Botterud [Bot03], Fleten and Näsäkkälä [FN03], Jensen and Meibom [JM08], Klein et al. [Kle+08]). Thereby, two different approaches concerning the modeling of electricity prices used for the power plant valuation can be distinguished. The first one assumes that electricity generation companies are price-takers [Den99; Den05; FN03]. This assumption has the advantage that the electricity prices can be directly modeled as a stochastic process. The other approach argues that the construction of new capacity has an impact on electricity prices, and therefore the assumption of a price-taker is inappropriate. In this case, a fundamental electricity market model is used to derive electricity prices [Bot03; JM08; Kle+08]. Especially if investments in several units should be considered, these investments influence electricity prices, and the latter approach is the appropriate one in our opinion.

Besides fuel and electricity price uncertainty, the changing legislation introduces additional challenges to power generation companies. In recent years, two political decisions with a significant impact on the development of the German electricity market have been taken: the nuclear phase-out and the introduction of the European Union emission trading system (EU ETS). The nuclear phase-out—if it is not delayed or scrapped by the new government—increases the demand for new generation capacity. The EU ETS obliges the generation companies to get an emission allowance for each ton of CO_2 they emit. In the past years, these emission allowances have been allocated for free, but in the future, emission allowances will probably, at least partly, be auctioned. Hence, the EU ETS introduced an additional source of uncertainty, which must be taken into account when power plant investments are planned: uncertainty about future carbon prices.

Aim of this Thesis

The objective of this thesis is to develop models for power plant investment planning that allow an adequate representation of fuel price and carbon price risks. Thereby, we take the perspective of a single generation company. This implies that investments of competing companies have to be modeled. As investments in new units influence the profitability of the existing ones, the existing portfolio of the considered company should be taken into account.

The models that are developed in this thesis should consider the interdependence between fuel prices, carbon prices and the existing power generation fleet on the one side and electricity production and electricity prices on the other side. While the models developed in this thesis should be applicable to different electricity markets, we only consider the German market in this thesis. For most power plant projects in Germany, two alternatives are currently discussed: coal-fired units and CCGT units. Hence, we focus on these two alternatives in this thesis.

Methods

We apply the real options approach to model power plant investments. Thereby, we use a fundamental electricity market model to derive the contribution margins for new units. Models combining the real options approach with fundamental electricity market models exist already. However, the existing models suffer from one problem: stochastic dynamic programming (SDP), the method used to solve the investment problem, is only suited to solve low dimensional problems. For that reason, fuel price and carbon price uncertainty is disregarded or only a few different scenarios are considered.

In this thesis, we aim to apply state-of-the art methods from operations research to be able to solve the power plant investment problem considering a wide variety of different fuel and carbon price scenarios. For this purpose, we use two methods, which have—to the best of the author's knowledge—not yet been applied to optimize power plant investments: SDP based on an adaptive grid method, and approximate dynamic programming (ADP).

The first method is directly based on SDP. However, to reduce the complexity of the problem, an approximated problem consisting of a subset of the state space is solved instead of the original problem. The subset of the state space that is used for the approximated problem is obtained by placing a grid in the state space. The approximated problem is then iteratively solved and the grid is adaptively refined by adding new states. To approximate the value of states that do not belong to the grid, but which may be reached during the optimization, we use an interpolation method. For this reason, we refer to this approach as interpolation-based stochastic dynamic programming (ISDP).

While ISDP allows to significantly reduce the number of considered states, it still requires several simplifications to solve the power generation expansion problem under fuel and carbon price uncertainty. For example, to be able to consider construction lead-times of a new plant, we allow investments only every four years. ADP promises to circumvent these restrictions. Instead of explicitly saving the value of each state in the problem, ADP approximates the value of a state, e.g., based on function approximations. This approach has two advantages: first, it significantly reduces the memory requirements, and second, it avoids an enumeration of the complete state space.

Combining these two methods with a detailed modeling of the electricity market, we hope to provide a methodology suited to optimize power plant investments under fuel and carbon price uncertainty. Thereby, the focus of our model lies on a strategic level. This leads to several simplifications. For example, we do not consider the location problem. In our fundamental electricity market model, we disregard the transmission network. We do not address the question how the investments are financed—we simply assume that the considered operator can borrow an infinite amount of money at a fixed interest rate. While we consider the existing portfolio of the investor, we do not explicitly model decommissioning decisions of old units.

Thesis Outline

This thesis is organized as follows: In Chapter 1, we give an introduction into power generation expansion planning. We start with a description of different electricity markets and some characteristics of electricity prices. We outline how electricity prices are derived in a competitive market, and we discuss the uncertainties to which power plant investments are exposed. We describe how energy prices are modeled and present different methodologies used to analyze and valuate power plant investments. We conclude the chapter with a summary of related work.

Chapter 2 is devoted to the modeling of electricity prices with fundamental electricity market models. We focus on two models with different levels of detail. The first model, based on the load–duration curve, is relatively simple. Due to its fast running times, it is well suited to calculate the contribution margins of new units considering many different scenarios. The second model is based on a detailed description of the electricity market as a linear programming model. To reduce the complexity of such a model, several simplifications like aggregation of similar units or the restriction to some typical days are generally made. We discuss several issues with these simplifications, which are often overlooked in literature. Then, we compare the results of the presented models with observed electricity spot prices.

Chapter 3 describes our ISDP investment model. Based on a Monte Carlo simulation of fuel and carbon prices, we select many different representative scenarios for which we calculate the contribution margins with one of the fundamental electricity market models. To reduce the computational burden, we start with an approximated problem based on a coarse grid, which is iteratively refined. We compare several criteria used to select the regions in the state space that are refined. Thereby, two different approaches can be distinguished: one approach tries to reduce the approximation error, the other one focuses on refinements around the optimal policy. Our tests show that a combination of both of these approaches works best. Finally, we compare the ISDP approach with a normal SDP approach and a deterministic approach based on expected fuel and carbon prices.

In Chapter 4, we use ADP to model power plant investments. We assume that the value of a state can be described by a function depending on the number of new units. For this purpose, we propose piecewise-linear value function approximations. This approach significantly reduces the computational burden compared to the ISDP approach and allows us to consider annual investments. In the tests we do, the ADP approach performs slightly better than the ISDP model. However, the algorithm turns out to be very sensitive to parameter choices, and has the further disadvantage that it cannot be extended to consider additional investment alternatives.

In Chapter 5, we perform several case studies based on the ISDP model. Thereby, we examine the effects of different assumptions used in our model, for example, the stochastic process used to simulate fuel and carbon prices. This case study yields a counterintuitive result: the higher the fuel prices and the uncertainty, the more units are built on average. In addition, we analyze the effect of different scenarios concerning the nuclear phase-out and the emission allowance allocation scheme. We show that a delay of the nuclear phase-out will reduce the incentives for investments in new units. Furthermore, we demonstrate that the current fuel-dependent allocation of emission allowances favors coal units. When coal units and CCGT units get the same amount of emission allowances, CCGT units are preferred. In addition, we evaluate the profitability of coal units with the carbon capture and storage technology. Our tests indicate that this technology will replace coal units without carbon capture technology.

In Chapter 6, the last part of this thesis, we give a conclusion and possible directions of future research.

Chapter 1

Introduction

In this chapter, we give an introduction into electricity markets and provide an overview of different methods applied to power plant valuation. We start with a description of electricity as a commodity. We introduce different markets where electricity is traded, namely the spot market, the derivatives market, the reserve market and the capacity market. Next, we discuss some peculiarities of electricity compared to other commodities: the non-storability of electricity, the fact that supply must always match demand, the price-inelastic demand and the grid dependency. These peculiarities are the reason for some characteristics of electricity spot prices like high volatility and extreme price peaks.

In the next section, we describe how electricity prices are determined. For this purpose, we introduce the microeconomic concept of pricing under perfect competition. Then, we discuss whether the assumption of perfect competition is justified in deregulated electricity markets. While it is questionable whether electricity markets are fully competitive nowadays, we argue that the principle of price determination under perfect competition can be used to derive electricity prices for power plant valuation methods.

In the following section, we focus on the uncertainties to which power plant investments are exposed. These uncertainties can be classified into volume risk, price risk and cost risk. We discuss which factors influence these risks, and how they are related. An important source of uncertainty are future fuel, carbon, and electricity prices. We provide an overview how these prices can be modeled.

Afterwards, we introduce the two most popular methods for valuating power plant investments, the levelized costs of electricity production approach and the real options approach. While the first one was mainly used in a regulated market and does not take into account uncertainties, the latter one allows a more appropriate consideration of uncertainties and is therefore better suited in a deregulated market. Finally, we give an overview of the research done so far in the field of power generation expansion models. Thereby, we describe simulation models, game-theoretic models and optimization models.

1.1   Electricity as a Commodity

From an economical point of view, electricity is a commodity. A commodity is some good for which there is demand, but which is supplied without qualitative differentiation across a market (O'Sullivan and Sheffrin [OS03, p. 152]). Following this definition, commodities produced by different producers are considered equivalent. Most of the commodities are basic resources and agricultural products such as iron ore, crude oil, coal, sugar or wheat. These commodities are traded at spot and derivatives markets, like the New York Mercantile Exchange. The price of a commodity is determined as a function of its market as a whole. However, due to transportation restrictions and costs, regional prices may be different.

In this section, we describe electricity as a commodity. We first introduce different markets where electricity is traded. Then, we discuss some peculiarities of electricity compared to other commodities. These peculiarities, e.g., the non-storability of electricity, explain why spot market electricity prices differ significantly from spot prices of other commodities. Finally, we briefly discuss why electricity derivative prices do not share these characteristics of electricity spot prices.

1.1.1   Electricity Markets

Like other commodities, electricity is traded on commodity exchanges. One of the most important power exchanges in Europe is the German power exchange, the European Energy Exchange (EEX) in Leipzig. The EEX consists of a spot market and a derivatives market. On the spot market, electricity, natural gas and emission allowances are traded. On the derivatives market, also coal futures are traded. In May 2009, the EEX and the Paris based Powernext decided to cooperate. The spot market activities of both exchanges are now concentrated on the Paris based EPEX Spot SE, while the derivatives are traded at the EEX Power Derivatives in Leipzig. Both markets include France, Germany, Austria and Switzerland.

	2002	2003	2004	2005	2006	2007	2008
Spot market	31	49	60	86	89	124	154
Derivatives market	119	342	338	517	1,044	1,150	1,165

Table 1.1: Annual traded volumes on the EEX spot and derivatives market for electricity in TWh. Source: EEX [EEX08]

The importance of the energy exchanges has steadily increased, as it can be seen looking at the traded volumes shown in Table 1.1.[1] The volumes traded on the derivatives market are much higher than those on the spot market. In total, the traded volumes at the EEX in 2008 have been about twice the annual electricity demand of Germany.

In the following, we describe four different markets on which power plants can generate revenues. Two of them, the electricity spot market and the derivatives market, are used to sell electricity. On the reserve market, revenues are obtained for providing reserve electricity that might be activated in case of short-term deviations between supply and demand. The forth market, the capacity market, does not exist in Germany. We describe it nevertheless, as it exists in other countries (e.g., in the U.S., in Spain, in Italy), and it might be discussed in Germany in case that the current framework does not provide enough incentives for investments in new units.

- **Spot market:**

  The spot market is composed of two different markets: a day-ahead market and an intra-day market. On the day-ahead market, which exists since 2002, electricity is traded for the next day. Electricity may be traded for individual hours or for either predefined blocks (e.g., base load, peak load[2]) or user-defined blocks. Bids for the next day must be submitted until 12.00 am, and must be between $-3,000$ €/MWh and $3,000$ €/MWh.

  To be able to react on short-term deviations from the expected schedule, the EEX created an intra-day market in September 2006. The intra-day market allows the trading of hourly electricity contracts or predefined blocks of electricity contracts for the actual and the next day up to 75 minutes before the

---

[1] It should be mentioned that a part of the increase is due to the extension of the market to Austria (April 2005) and Switzerland (December 2006).

[2] The base-load period comprises all hours of a day, while the peak-load period comprises the hours from 8.00 am to 8.00 pm.

start of the delivery. The trading for the next day starts at 3.00 pm. The allowed price span is from $-9,999$ €/MWh to $9,999$ €/MWh. Similar to the day-ahead market, the volumes traded on the intra-day market have significantly increased during the last years. During the first half year of 2007, the average traded daily volume was 2.7 GWh. During the same period in 2009, it has risen to 10.6 GWh. However, compared to the total volume traded in the spot market, these values correspond to less than 3%.

The spot market allows electricity generation companies to optimize their short-term unit commitment. Based on the spot prices, generation companies may decide to commit additional units and sell their production, or they may decide not to commit a unit and buy the required electricity at the spot market.

- **Derivatives market:**
On the derivatives market, two different types of derivatives are traded: options and futures. An option gives the buyer the right to buy (call option) or sell (put option) an underlying security for the specified price on the last day of trading (European option) or until the last day of trading (American option). A future on the contrary is an obligation to buy or sell an underlying security for the specified price at a certain point of time in the future. On the EEX, base load and peak load futures with different maturities can be traded. The delivery periods for these futures are a month, a quarter year or a year. Monthly futures are traded up to nine months in the future, quarterly futures up to eleven quarters in the future and yearly futures up to six years in the future. Additionally to these futures, options on the base load futures for the three different delivery periods are traded. Options can be traded up to the next three years for the yearly futures.

The derivatives market allows electricity generation companies to reduce their price risk by selling long-term contracts. The same is true for electricity trading companies, which can purchase electricity for longer time periods. This increases their planning security and allows them for example to offer price guarantees to their customers.

- **Reserve market:**
One of the peculiarities of electricity described in the next subsection is the fact that supply must always match demand. It is the task of the transmission system operators (TSOs) to ensure that this condition is fulfilled. To

be able to achieve this, reserve energy is used to react on short-term differences between supply and demand. Three forms of reserve energy are distinguished in Germany: primary control power, secondary control power and minute reserves. Primary control power must be activated within 30 seconds, and can be used up to 15 minutes. Secondary control power must be supplied within five minutes, while minute reserves must be activated within 15 minutes and can be used up to one hour.

In Germany, the reserve market is organized by the four TSOs (50Hertz Transmission, Amprion, EnBW Transportnetze, transpower) over the Internet platform www.regelleistung.net. Minute reserves are procured via daily tenders, primary and secondary control power are traded on a monthly basis. Primary control power comprises positive and negative control power, while for secondary control power and minute reserves bids can be separated for negative and positive reserves. Bids for primary control power consist of one price, the price for the supply of control power, independent of its usage. Secondary control power and minute reserve bids contain in addition a price for the usage of the reserve energy.

The reserve market is particularly interesting for generation companies with pumped-storages and fast starting peak-load units, since already the supply of reserve capacity is paid. Even though peak-load units often have high costs for electricity generation, they may be selected to provide reserve power, as the bids for reserve power consist of two prices, a capacity price and an energy price. The participation in the reserve market allows these units to generate some income even without actually being used to produce electricity.

- **Capacity market:**
  A controversial debate exists whether liberalized electricity markets are able to provide enough investment incentives to ensure a long-term security of supply. Especially peak-load units, which run only a few hours per year, may not be able to recover investment costs. Some countries have tackled this problem by introducing capacity payments or capacity markets.

  A straightforward way to provide investment incentives is to give generators a per-MW payment for available capacity (independent of production) or as an adder to the price for generated electricity. Such capacity payments have been introduced in the UK, Spain, Argentina and Italy for example.

However, in the UK, capacity payments have been dropped in 2001. Details about the implementation of capacity payments in the before mentioned countries are described for example by Perekhodtsev and Blumsack [PB09].

Instead of providing a capacity payment, another approach to ensure an adequate level of supply is to fix the required amount of capacity. Load serving entities must guarantee a prescribed level of reserve capacity above their peak load within a certain time frame. Capacity obligations can be exchanged on a capacity market. Such capacity markets have been established in the United States, e.g., in New York, New England and the Pennsylvania–New Jersey–Maryland Interconnection (PJM) (for a detailed description of capacity markets, see, e.g., Cramton and Stoft [CS05]).

There is no consensus whether capacity payments or capacity markets provide efficient investment incentives or not. Consequently, some countries have implemented some sort of capacity remuneration, other countries—including Germany—have not. The discussion of the most efficient market design to ensure adequate levels of investment is beyond the scope of this thesis. For some arguments against capacity payments or capacity obligations, see Oren [Ore05]. Arguments for the need of capacity markets are presented, e.g., by Joskow [Jos08].

1.1.2 Peculiarities of Electricity

While electricity is a commodity, it is characterized by some peculiarities, which are important to consider since they have significant effects on electricity prices. These peculiarities are the non-storability of electricity, the fact that supply must match demand anytime, the (short-term) price-inelasticity of the electricity demand and its grid-dependency.

- **Non-storability:**
  An important characteristic of electricity is its non-storability. Electrical energy may be transformed to another form of energy to be stored. However, a significant amount of energy is lost during such a transformation. There are mainly two different alternatives to store a greater amounts of energy: pumped hydro-storages and compressed air energy storages (CAES). Pumped hydro-storages use electricity to pump water from a lower reservoir into an upper reservoir. When additional electricity is required, water from the upper reservoir is released through a turbine. Pumped hydro-storages

are by far the most important storage technology with a total capacity of about 6,610 MW installed in Germany (Konstantin [Kon08, p. 263]). The average round trip efficiency[3] of pumped hydro-storages is between 70–85%. Owing to the requirements of pumped hydro-storage locations, the potential for additional storages is very limited in Germany.

CAES use electricity to compress air and save it in subterranean caverns or porous rock layers. To produce electricity, the compressed air is used to drive a generator via a turbine. Only two CAES exist worldwide. One of those was built 1978 in Huntdorf (Germany) and has a capacity of 290 MW. The overall efficiency of this CAES is 42% (Grimm [Gri07, p. 13]).[4]

Even though pumped hydro-storages and CAES provide a limited possibility to store energy that can be transformed to electricity, the principal non-storability of electricity has significant effects on electricity prices. This is mainly due to the second peculiarity of electricity, the requirement that supply must match demand at every moment in time.

- **Supply must always match demand:**
  If electricity supply does not match electricity demand, the line frequency changes. A surplus of supply leads to an increased frequency, a shortage of supply to a decreased frequency. While minor fluctuations of the line frequency cannot be avoided, exceptional or rapidly changing line frequency leads to power outages. An example for such an outage could be observed on November 4, 2006. Over five million people in Germany, France, Italy, Belgium and Spain were affected by a breakdown. The outage occurred due to power imbalance caused by a shutdown of several transmission lines (UCTE [UCT07]). The significance of the obligation that supply must always match demand is increased by the price-inelasticity of electricity demand.

- **Price-inelastic demand:**
  Short-term electricity demand exhibits very low short-term price elasticity (Lijesen [Lij07]). This low price elasticity can be explained by the fact that most electricity companies charge their customers based on an average cost tariff. Due to the missing price signal, consumers have no incentive to shift demand from high price hours to low price hours.

---

[3]The round trip efficiency is the overall efficiency for converting electricity into another form of energy and back to electricity.

[4]The second storage with a capacity of 110 MW is located in McIntosh, Alabama (United States)

In recent times, the question how the demand side can improve the efficiency of electricity markets has attracted a lot of attention. A precondition
for an active demand side is smart metering. Smart metering identifies electricity consumption in more detail than a conventional meter. Using smart
metering, price signals can be passed to consumers, to give them incentives
to adjust their electricity demand based on the current prices. Even though
smart metering has the potential to increase the (short-term) price elasticity of electricity demand, many challenges remain. For example, it is not
clear whether the benefits of smart metering outweigh its costs (Haney et
al. [Han+09]).

Smart metering may contribute to a more efficient electricity production in
the future by providing the basis for a price elastic demand. This would
reduce the volatility of electricity prices, as it shifts some demand from peak-
load hours to periods of lower demand. However, at the moment it still
seems reasonable to model electricity demand as (almost) price-inelastic in
the short term.

- **Grid dependency:**
  Finally, another important characteristic of electricity is its grid-dependency.
  As a consequence, there exist several separated markets for electricity in Europe. These markets mainly correspond to the national territories. While
  the national electricity grids are coupled, the connections between different
  countries are often a bottleneck in the transmission system. Therefore, different prices can be observed on different markets.

  Such a case can be seen in Figure 1.1. The figure shows the day-ahead electricity prices on June 18, 2009 for Germany, Switzerland, France and Denmark. While the prices for Germany, Switzerland and France are very similar, the prices in Denmark are considerably lower during the high price periods. This is an indicator for a congested transmission line between Denmark
  and Germany. Otherwise, there would be an arbitrage opportunity by buy-
  ing electricity in Denmark and selling it in Germany.

1.1.3   Spot Market Electricity Prices

Spot market electricity prices show some remarkable characteristics, which are
mainly due to the peculiarities of electricity discussed in the previous subsection
(cf. Geman and Roncoroni [GR06]). These characteristics of electricity prices are

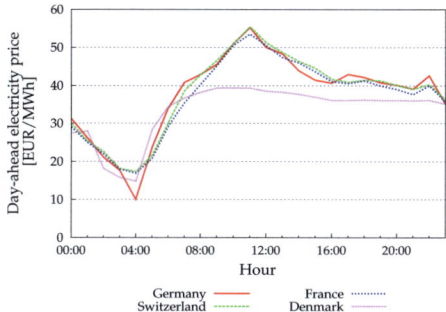

Figure 1.1: Day-ahead electricity prices for June 18, 2009 in Germany, Switzerland, France and Denmark. Sources: EPEX Spot, NordPool

cycles, a high volatility combined with extreme price peaks, a mean-reverting behavior, and the occurrence of negative prices.

- **Cycles:**

  Electricity prices show daily, weekly and seasonal cycles. During the low-demand hours of the night, electricity is relatively cheap. In times of higher demand, electricity gets more expensive. On a summer day, electricity prices usually reach their maximum at noon, then they start to decrease. On a typical winter day, an additional peak is reached during the evening hours, as it can be seen in Figure 1.2(a).

  The second characteristic pattern for electricity prices is the weekly cycle. While prices show a similar curve for each weekday, they are different on the weekend. First, the price level is generally lower. Second, the difference between the highest and the lowest price is generally much lower than during weekdays (see Figure 1.2(b)). The reason for lower prices on weekends is the lower demand, mainly due to reduced demand from business activities.

  In addition to these daily and weekly cycles, also seasonal cycles can be observed. These cycles are caused by temperature differences and the number of daylight hours. In hydro-dominated systems, electricity prices are additionally influenced by the availability of hydro-power.

- **High volatility:**

  Electricity prices exhibit the greatest volatility amongst commodity prices (see, e.g., Henning et al. [Hen+03] or Weron [Wer00]). However, as Benini et al. [Ben+02] remark, the volatility of electricity prices is less random than

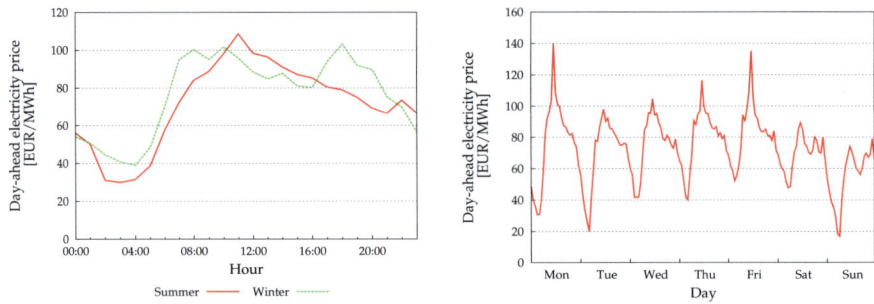

(a) Daily prices for a summer (July 16, 2008)   (b) Weekly prices (July 7, 2008–July 13, 2008)
and a winter (January 14, 2009) day

Figure 1.2: Typical day-ahead electricity prices. Source: EEX.

those of other commodities. This is due to the fact that a large part of the
electricity spot price variations is related to load variations, which can be
relatively well predicted. Furthermore, Benini et al. point out that volatility varies on different markets, a result which is also observed by Li and
Flynn [LF04]. The main reason for this effect is the structure and the size
of the power plant fleet in these markets. Markets with a high amount of
reserve capacity relative to the peak-load exhibit a lower variance. Also in
hydro-dominated systems such as the Scandinavian power market, volatility is much lower. This can be seen in Figure 1.3(a). It shows the hourly
day-ahead prices for June 2009 at the EEX (Germany and Austria) and the
NordPool (Scandinavia) power exchange. While both price curves show the
characteristic daily and weekly cycles, it is noticeable that the amplitude of
EEX prices is much greater than those of NordPool prices.

- **Spikes:**
  In addition to their high volatility, electricity spot market prices are characterized by high price spikes. An extreme example of such a spike could be
  observed at the EEX on July 25, 2006. Day-ahead electricity prices rose from
  55.99 €/MWh at 6.00 am to 2,000.07 €/MWh at 11.00 am (see Figure 1.3(b)).
  These extremely high prices were caused by outages of some nuclear, lignite
  and coal plants due to the lack of cooling water caused by the extreme heat.
  Two days later, another abnormally high peak was reached. Two additional
  days later, the spot prices reached their normal level.

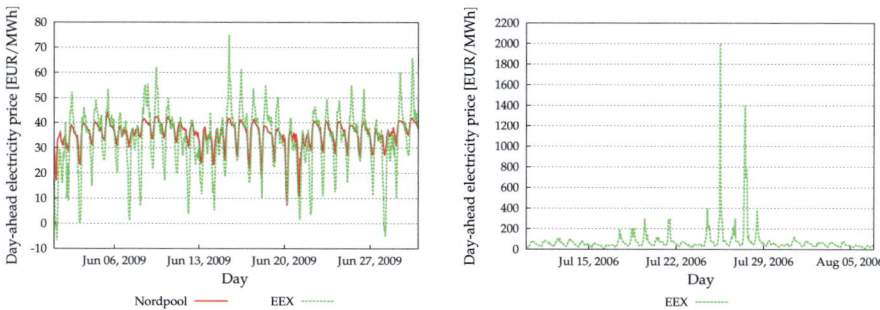

(a) Day-ahead electricity prices for June 2009 at the EEX and NordPool.

(b) EEX day-ahead electricity prices from July 10, 2006 to August 6, 2006.

Figure 1.3: Volatility and spikes in electricity spot prices. Sources: EEX, NordPool.

- **Mean-reversion:**

  It is generally assumed that electricity prices show a mean-reverting behavior (see, e.g., Deng [Den99]). This behavior is due to the way electricity prices are determined. In a competitive market, it is assumed that electricity prices correspond to the marginal costs of the most expensive unit producing electricity. A forced plant outage may change the power plant that sets the price. As a consequence, electricity prices rise. When the outage is resolved, prices return to the normal level.

  Such a behavior can be be seen in Figure 1.3(b). At the beginning of the plotted period, spot prices are on their normal level. Starting with the July 17, 2006, spot price peaks during the weekdays are significantly higher than normal. This level is maintained for two weeks with extreme peaks on the July 25 and July 27. After the reason for peaks, the forced plant outages, disappeared, prices reverted back to their normal level.

- **Negative prices:**

  Unlike other commodities, electricity prices can become negative. This phenomena is caused by two reasons. The first one is the fact that supply must always match demand. The second reason are the technical constraints of power plants.

  In case of unforeseen short-term deviations from the expected demand or from the feed-in of electricity produced by intermittent sources like wind power, electricity producers must immediately adjust their production. If there is an unforeseen high amount of wind power in the network, the feed-in from conventional power plants must be reduced. However, most thermal units have a minimum production level. If the units of an operator are already producing at their lower output level, the operator may has to shut down a plant. Shutting down a power plant has further consequences. Once a power plant is shut down, it (generally) has to stay offline for a certain time. Additionally, starting-up the plant later on causes additional start-up cost due to abrasion and additional fuel usage. To avoid these additional costs, it may be cheaper to pay some fee for producing electricity instead of shutting down a power plant.

Negative prices for intra-day trading are allowed on the EEX since December 19, 2007. Negative prices could be observed for example on December 22, 2008. Between 1.00 am and 5.00 am, the intra-day electricity prices have been below $-20 \, €/MWh$.

1.1.4   Electricity Derivative Prices

The characteristics of electricity prices discussed so far are typical for spot prices. While derivative prices are generally at least partly explained by the current spot prices, they often behave different.

A common concept in pricing derivatives of commodities is the so called convenience yield. It was first introduced by Kaldor [Kal39]. The convenience yield is the additional value an owner of a physical commodity has compared to someone who owns a future contract on the same commodity. This additional value is the possibility to use the commodity when it is wanted. An example for this additional value can be seen in the case of shortage of the commodity. In such a case, it is beneficial to own the commodity to keep production going instead of just owning a futures contract. The convenience yield concept plays an important role in most models for commodity prices (e.g., Brennan and Schwartz [BS85], Pindyck [Pin01], Schwartz [Sch97]). However, given the non-storability of electricity, the convenience yield concept is not applicable to electricity derivatives.

Bessembinder and Lemmon [BL02] propose that the forward power price will generally be a biased forecast of the future spot price, with the forward premium decreased by the anticipated variance of wholesale spot prices and increased by the anticipated skewness of wholesale spot prices.

As the delivery period of the electricity derivatives traded at the EEX is at least a month, the derivative electricity prices can be seen as average expected future spot prices. Therefore, they are much less marked by the characteristics typical for spot prices. The volatility as well as the degree of mean-reversion are much lower. This is also due to the fact that—other than electricity itself—electricity derivatives can be stored and sold later.

1.2 Electricity Pricing Theory

An essential input for any power plant valuation model are electricity prices, as the revenues generated by power plants strongly depend on them. In this section, we give an overview how electricity prices are determined.

We start with an introduction to microeconomic pricing theory. Depending on the market structure, different models are used to determine prices. The different types of market structures are perfect competition, oligopoly and monopoly. Most fundamental electricity market models, like the ones we use in this thesis, are based on the assumption of perfect competition. Therefore, we present in the following the microeconomic theory of price determination under perfect competition. A description of price determination in an oligopoly or a monopoly can be found in most microeconomic textbooks, e.g., Varian [Var06].

In the second part of this section, we discuss whether the assumption of perfect competition, on which most fundamental electricity market models are based, is pertinent. While there is evidence that the German electricity market does not correspond to a fully competitive market at the moment, it is assumed that the degree of competition will increase further. Hence, we argue that the assumption of perfect competition can be used for the determination of electricity prices.

1.2.1 Price Determination under Perfect Competition

A market with perfect competition is characterized by many buyers and sellers and a homogeneous good. It is assumed that neither a single buyer nor a single seller can influence the price. Both, sellers and buyers are therefore modeled as "price-takers". Every seller can specify his output quantity q as a function of the market price p. This is the (individual) supply function $Q_s(p)$. The same function, rewritten as a function of the quantity instead of the price, $P_s(q)$, is called inverse

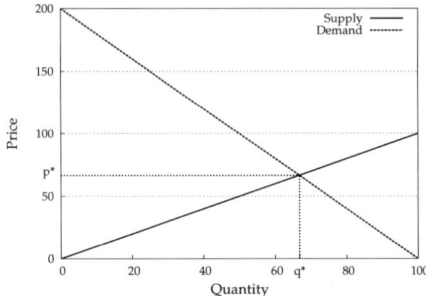

Figure 1.4: Price and quantity determination in a competitive market.

supply function.[5] As we will see later, the inverse supply function—which is also called the merit-order curve in electricity markets—is often used to explain how electricity prices are determined. The demand function $Q_d(p)$ expresses the buyers demand q as a function of the price p. Similar to the supply function, also the demand function may be written as the inverse demand function, $P_d(q)$. Combining the individual supply functions and the individual demand functions leads to the aggregate demand functions and the aggregate supply functions, respectively. The equilibrium market price p^* is determined by the intersection of the aggregated demand curve and the aggregated supply curve, $Q_d(p^*) = Q_s(p^*)$.

The reason why this condition must hold in an equilibrium can be seen by considering what happens if it is not satisfied. If a price $p' < p^*$ is observed, and demand is greater than supply, some sellers will notice that they can get a higher price. If enough sellers will notice this, they will all adjust their prices, until the equilibrium price is reached. If $p' > p^*$ and the demand is lower than the supply, the suppliers cannot sell the calculated quantities. The only possibility to sell more is to lower the price. If one supplier lowers the price, the other suppliers have to follow him, as the good that is sold is assumed to be homogeneous.

Example. *Figure 1.4 shows an example for the inverse aggregate demand function and the inverse aggregate supply function. The inverse aggregate supply function is $Q_s(p) = p$, while the inverse aggregate demand is $Q_d(p) = 100 - \frac{1}{2} \cdot p$. The equilibrium price p^* and the equilibrium quantity q^* can be determined as the intersection of both curves. In this example, the equilibrium price is 66.66, the equilibrium quantity is 66.66.*

Under perfect competition, it is assumed that sellers bid corresponding to their marginal costs. This result can be derived considering the profit maximization

---

[5]Economists often refer to both functions as supply function.

problem of a single producer. The profit Π is equal to:

$$\Pi = \sum_{i=1}^{n} (p_i \cdot y_i) - \sum_{i=1}^{m} (w_i \cdot x_i).$$

The first term describes the revenue of the outputs y_i, which can be sold for prices p_i. The second term denotes the costs associated with the production of the outputs. The required input factors are denoted with x_i, the corresponding prices of these input factors are w_i.

Consider the simple case with one output product, for which two different input factors are needed. Assume that the quantity of input factor two is fixed at $\bar{x}_2$. This factor might represent the capital costs for the machines used for the production, or a rental fee. Let $f(x_1, x_2)$ be the production function specifying the output quantities depending on the input parameters. The profit maximization problem is then

$$\max_{x_1} \Pi = f(x_1, \bar{x}_2) \cdot p - w_1 \cdot x_1 - w_2 \cdot \bar{x}_2.$$

As a necessary condition for an optimum, the derivative of the profit function must be equal to zero. Therefore, the following condition must be satisfied for the optimal solution x_1^*:

$$p \cdot MP_1(x_1^*, \bar{x}_2) = w_1 \tag{1.1}$$

with MP_1 being the marginal product[6] of input factor one. The marginal product of a input factor multiplied by the price of the output factor must correspond to the costs of the input factor. If the marginal product multiplied with the output price is higher than its costs, it would be profitable to produce more. If the marginal product multiplied with the output price is lower than the costs of the input factor, then profits can be increased by producing less. Dividing equation (1.1) by $MP_1(x_1^*, \bar{x}_2)$, we get

$$p = \frac{w_1}{MP_1(x_1^*, \bar{x}_2)} = MC_1(x_1^*, \bar{x}_2),$$

where $MC_1(x_1^*, \bar{x}_2)$ denotes the marginal costs of the input factor one. It is important to note here that the optimal solution is independent of the fixed costs.

This result implies that—under the assumption of perfect competition—electricity prices are equal to short-term marginal costs. Capital costs and other fixed costs are irrelevant for the decision whether to produce electricity or not. However, this

---

[6]The marginal product is the extra output produced by one more unit of an input.

result holds only under the assumption of perfect competition. In the following subsection, we discuss whether this assumption is pertinent.

1.2.2   Price Determination in a Deregulated Electricity Market

After the introduction to microeconomic pricing theory, we examine the price formation in a deregulated electricity market. In a deregulated electricity market, prices are determined by the bids of the suppliers and the consumers in the market. Thereby, the bidding strategy of suppliers depends strongly on the degree of competition. Under perfect competition, profit maximizing bids are equal to marginal costs, while a decreasing decree of competition leads to increasing profit maximizing bids. As electricity is a homogeneous good, there is only one market-clearing price. Under perfect competition, this market clearing price corresponds to the marginal costs of the most expensive power plant required to satisfy the demand.

A central idea of the electricity market liberalization has been to increase the degree of competition. However, it is questionable whether observed electricity prices correspond to those of a competitive market. There are several studies comparing observed electricity prices in Germany with the assumed short-run marginal costs. We summarize these studies in the following, as the results are important for the question whether the assumption of perfect competition used by most fundamental electricity market models is pertinent or not.

Müsgens [Müs06] studies the degree of market power in the German electricity market based on a fundamental electricity market model. The model consists of several regions and considers technical restrictions such as plant capacities or start-up costs. Following the theory of pricing under perfect competition, the marginal costs of the most expensive unit producing electricity are assumed to correspond to the electricity prices. The prices derived by the model are compared to the spot market electricity prices observed at the EEX. Differences between model prices and observed prices are attributed to the exercise of market power. For the period of June 2000 to August 2001, no exercise of market power can be observed. For the period from September 2001 to June 2003, Müsgens reports strong evidence for the exercise of market power with average prices being nearly 50% above estimated costs. Most of the price differences occur during high-load periods.

Lang [Lan08a] uses a combination of a mixed-integer and a linear programming fundamental electricity market model to study the German electricity spot prices

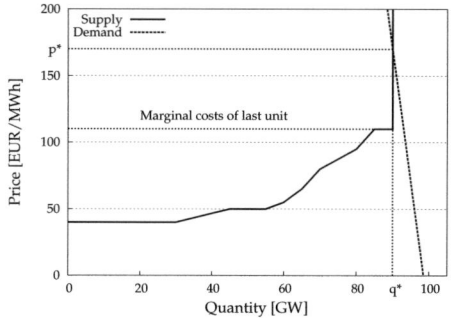

Figure 1.5: Price formation in case of scarcity.

between 2000 and 2005. Lang reports that in 2003, 9.7 €/MWh of the spot prices can be attributed to the exertion of market power. In 2005, market power is assumed to be responsible for 8.0 €/MWh.

Hirschhausen et al. [Hh+07] compare EEX spot prices between 2004 and 2006 with marginal production costs based on a merit-order curve. Hirschhausen et al. report significant mark-ups on realized prices compared to marginal costs, especially for mid load and peak load segments.

However, there are also other opinions. In a study commissioned by the *VRE* (the association of Germany Electricity Producers), Swider et al. [Swi+07] argue that the conclusions of the before mentioned studies, the exertion of market power, do not hold. Swider et al. remark that an important problem of these studies is data uncertainty, especially concerning the real power plant availabilities and fuel prices. An example from [Hh+07] is used to demonstrate that the deviations between marginal costs and electricity prices observed by Hirschhausen et al. can also be explained by data uncertainty. Furthermore, Swider et al. point out that electricity prices solely based on short-run marginal costs are not sufficient to cover investment expenditures. Therefore, Swider et al. conclude that it is natural to observe sometimes higher prices, which allow to recover long-run marginal costs.

Ockenfels [Ock07] also argues that electricity prices must sometimes be over short-term marginal costs to provide sufficient incentives for investments. He points out that this is in line with the theory of pricing under perfect competition. Figure 1.5 illustrates such a case. In case of scarcity, when demand is higher than the available capacity, the market clearing price is above the marginal costs of the most expensive unit. The price is then determined by the demand. If the price was corresponding to the marginal costs of the most expensive unit, not all demand could be satisfied. Therefore, the price is increased until the demand matches the maximum supply. This price is called the *Value of Lost Load*(VoLL). However, under perfect competition, this situation can only occur when demand exceeds the available capacity,[7] a situation which did not prevail in 2006 as it is mentioned by Weigt and Hirschhausen [WH08].

An explanation why price mark-ups are mainly observed in peak hours can be found considering the *Residual Supply Index (RSI)* (see, e.g., Lang [Lan08b]). The RSI is defined as $RSI = (\sum_{i=1}^{n} C_i - C_x) / Load$, where $\sum_{i=1}^{n} C_i$ is the capacity of all operators in the market, and C_x the capacity of the considered operator for whom the RSI is calculated. The RSI quantifies the importance of the considered operator to satisfy the demand. The lower the RSI is, the more important is the considered operator. An RSI smaller than one signifies that the demand cannot be satisfied without the considered operator. It can be seen that the higher the load is, the lower is the RSI. This relationship explains why deviations between calculated marginal costs and observed prices mainly occur during high-peak hours. Lang [Lan08a] reports a small, but significant correlation between the RSI and the observed differences between marginal costs and electricity prices. He finds an RSI below one for E.ON in 2004 for 3,701 hours.

Similar values are reported by Möst and Genoese [MG09]. However, the authors point out that these values do not consider potential electricity imports. Considering the import potential, the number of hours with an RSI below one drops to 85 for E.ON in 2004. This is the maximum observed for the years of 2001 to 2006 in the German electricity market.

Summarizing the results of the different studies, it is questionable whether electricity markets are perfectly competitive nowadays. Nevertheless, electricity market models based on the assumption of perfect competition may be used to determine electricity prices. With a future extension of cross-border transmission capacity, market power will decline and competition will increase. Electricity markets will be closer to perfect competition as they are today. Under such circumstances, differences between observed electricity prices and marginal costs should also decrease.

---

[7]There is a second case, in which electricity prices may exceed marginal costs. This can occur, when the most expensive unit producing is at its maximum output level, and the marginal costs of the next expensive unit are higher. In this case, the price may be somewhere in between the marginal costs of the two units (see, e.g., Borenstein [Bor00]).

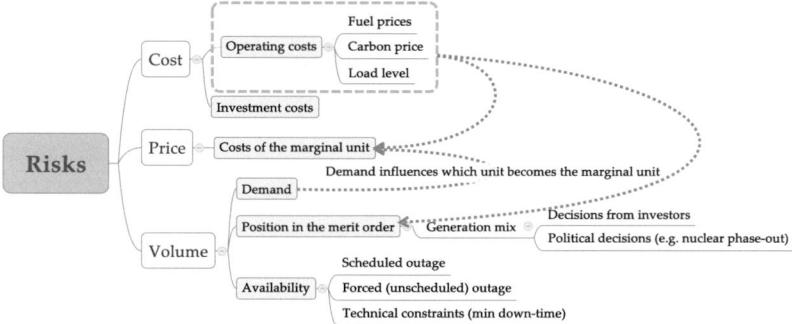

Figure 1.6: Uncertainties related to power plant investments.

Furthermore, electricity prices based on the assumption of perfect competition may be used as a lower bound for real electricity prices. Investments that are profitable under these prices remain profitable in case of higher electricity prices that may be observed due to the exertion of market power.

1.3 Uncertainties Related to Power Plant Investments

Power plant investments are heavily exposed to uncertainties. One reason for this is the relatively long planning horizon which must be considered when deciding on power plant investments. For example, coal units have an useful life of about 35 years (Konstantin [Kon08]). Another reason is the high volatility of fuel prices, which have an important influence on the generation costs.

Lemming [Lem05] proposes a classification of risks based on the cash flow generated by power plants, which is equal to the electricity production multiplied by the electricity price less the costs. Hence, Lemming distinguishes between cost, volume, and price risk. As shown in Figure 1.6, we use this classification, but go one step further in discussing the different sources of uncertainty and their impact on volume, price and costs. In our opinion, it is very important to understand these effects, and to consider these interdependencies in an investment model. In the following, we briefly describe the most important uncertainties related to power plant investments, and discuss how they are interrelated.

Thereby, we focus on the uncertainties influencing future contribution margins. There exist also other sources of uncertainty, e.g., about the exact investment costs or about the construction time of new units. However, we do not consider these uncertainties in our investment models, because these parameters are less uncertain than future contribution margins.

- **Fuel price uncertainty:**
  In Germany, about 55% of the electricity is produced by fossil fuels (coal, lignite and natural gas). The prices of these fuels, especially the one for natural gas, exhibit a strong volatility (as seen on Figure 1 on page xi). Depending on the type of power plant, fuel costs account for about 30% (for coal plants) and about 65% (for CCGT plants) of the total costs (including capital costs).[8]

  It is obvious that fuel price uncertainty has an important effect on generation costs. However, it also influences the electricity price. Under the assumption of perfect competition, the electricity price corresponds to the marginal costs of the most expensive unit producing electricity. As the price-setting unit is often either a gas or a coal plant, fluctuating fuel prices also influence electricity prices.

- **Electricity price uncertainty:**
  Spot market electricity prices are characterized by high volatility, spikes and cycles. There are different factors that influence electricity prices. Under the assumption that the electricity price depends on the costs of the marginal unit, these factors are the following ones: First, every uncertainty changing the costs of the marginal unit changes the price. Besides the fuel price uncertainty, this might be a change of the carbon costs. Second, a change of electricity demand or electricity supply can change the unit that sets the price. On the short-term, such an effect can be due to the stochastic feed-in from electricity produced by renewable sources (which must be integrated into the grid due to legal obligations). On the long-term, such changes might be due to a changing electricity demand or modifications of the power generation fleet. A change of the existing power generation fleet may be caused by future political decisions or the behavior of competitors, which are both uncertain.

- **Legal framework:**
  In the last years, there were two important political decision with a major impact on the electricity production in Germany. First, in 2002, a nuclear

---

[8]Based on the Levelized Costs of Electricity Production in OECD and IEA [OI05] for a discount rate of 10% and carbon prices of 20 €/t.

	Phase II	Phase III
Barclays	11–20	40
Cheuvreux	12–23	30
Orbeo	13.7–20	23–38

Table 1.2: Different carbon price forecasts [€/t] for Phase II (2008–2012) and Phase III (2013–2020) of the EU ETS. Source: Capoor and Ambrosi [CA09].

phase-out was decided.[9] This decision has an important impact on the structure of the future power generation fleet, and therefore also on future electricity prices.[10]

The second political decision having a major impact on the electricity generation was the implementation of an emission allowances trading system.[11] As a consequence, new costs were created, which have to be considered for power plant investments. Even though the generation companies have received the emission allowances for free in the past, several studies (e.g., Genoese et al. [Gen+07], Hirschhausen et al. [Hh+07]) have shown that carbon prices are passed through to electricity prices. Furthermore, as the CO_2 emissions depend mainly on the fuel, the position of the different units within the merit-order curve[12] now depends not only on the fuel prices, but also on the carbon prices. Future carbon prices are highly uncertain, amongst others because the amount of allowed emissions is only fixed for the period until 2012. As prices mainly depend on the amount of allowed emissions, price forecasts are difficult at the moment. This can be seen by comparing different studies, which result in quite different carbon prices as it can be seen in Table 1.2.

---

[9]"Vereinbarung zwischen der Bundesregierung und den Energieversorgungsunternehmen vom 14. Juni 2000"

[10]Political decisions are always subject to change. The new government plans to extend the operating life of nuclear power plants [CF09].

[11]Established by Directive 2003/87/EC

[12]The merit-order curve is the curve obtained by sorting all units from the least expensive to the most expensive unit and plotting the marginal costs as a function of the cumulated capacity (see Section 2.2 on page 67).

- **Electricity demand:**
  Changes in the electricity demand have two impacts. First, they may change the amount of electricity that is produced by the considered units. Secondly, by changing the marginal unit, electricity demand changes may also influence electricity prices.

- **Uncertainty about the behavior of competitors:**
  The decisions of competitors have a similar effect as changes in the electricity demand. On the one hand, additional units from competitors may decrease the amount of electricity produced by the own units. On the other hand, even if they have no impact on the electricity production of the own units, they may change the electricity price by changing the marginal unit.

1.4   Modeling Energy Prices

The most important uncertainties related to power plant investments concern the future contribution margins generated by new power plants. These contribution margins mainly depend on the development of fuel, carbon, and electricity prices. In this section, we discuss several possibilities to model these prices.

Energy prices are often modeled as a stochastic process. Hence, we start with a short description of two basic stochastic processes used for energy price modeling, the geometric Brownian motion (GBM) and the mean-reverting process. We continue with an overview of different methods of modeling fuel prices. Then, we briefly present approaches to model carbon prices. Finally, we examine how electricity prices can be modeled.

There are two different approaches to model electricity prices. They can either be modeled as a stochastic process, similar to the way fuel prices are commonly modeled. Alternatively, a fundamental electricity market model can be used to determine electricity prices depending on the current fuel prices, electricity demand and electricity supply. For our investment models, we opt for the latter approach, as it is better suited to consider the relationship between fuel prices, carbon prices, the existing power generation fleet on the one side and electricity prices and electricity production on the other side.

In the last two parts of this section, we give an overview of both approaches modeling electricity prices. We first consider the approach modeling electricity prices as a stochastic process, before we briefly outline different fundamental electricity market models. As the fundamental electricity market models are an important part of our investment models, we describe these models in detail in Chapter 2.

1.4.1 Stochastic Processes

In this subsection, we briefly describe two relatively simple stochastic processes used to model fuel prices: the GBM and the mean-reversion process. In spite of their relative simplicity—or maybe just because of it—they are frequently used to model fuel prices. Both processes are Markovian, which means that the next price depends only on the current price, while older prices are irrelevant for future prices. This property is a precondition for the application in an SDP model like our investment model.[13] A detailed discussion of the GBM, the mean-reverting process as well as other stochastic processes including the derivation of the formulas given in the following parts of this subsection is provided by Dixit and Pindyck [DP94].

1.4.1.1 Geometric Brownian Motion

A GBM is a continuous time stochastic process x_t that follows a stochastic differential equation of the form:

$$dx_t = \mu x_t dt + \sigma x_t dz_t,\tag{1.2}$$

where μ describes the drift of the stochastic process, σ the volatility, while dz_t are the increments of a Wiener process (also called Brownian motion) with

$$dz_t = \epsilon \sqrt{dt}.\tag{1.3}$$

Thereby, ϵ is a standard normally distributed random variable with an expected value of 0 and a variance of 1. In case of multivariate processes, the generated random numbers may be correlated.

For an arbitrarily chosen initial value S_0, equation (1.2) has the following analytical solution:

$$x_t = x_0 e^{(\mu - \frac{\sigma^2}{2})t + \sigma z_t},$$

---

[13]Non-Markovian processes may also be used in an SDP model. However, in such a case additional attributes must be added to a state. For a stochastic process depending on the last two observations, these two observations (instead of just the last one for a Markovian process) have to be saved for each state. This significantly increases the state space.

where x_t is a log-normally distributed random variable with an expected value $E(x_t) = x_0 e^{\mu t}$ and variance $Var(x_t) = e^{2\mu t} S_0^2 (e^{\sigma^2 t} - 1)$. One characteristic of a GBM is that, assuming a positive initial price, prices always remain positive.

The application of the GBM to model discrete price changes leads to

$$\frac{\Delta p}{p} = \mu \Delta t + \sigma \epsilon \sqrt{\Delta t}, \tag{1.4}$$

where p denotes the considered price (Weber [Web05, p. 41]). It can be seen from equation (1.4) that the relative price change $\Delta p / p$ depends on two terms: a deterministic drift $\mu \Delta t$ and a normally distributed stochastic variable with an expected value of 0 and a variance of $\sigma^2 \Delta t$. This deterministic drift implies that relative price changes depend only on time, but not on other factors like the current price level. The variance of the relative price changes grows with the time horizon. As a consequence, an increased time horizon leads to increased uncertainty in the predicted values.

1.4.1.2 Mean-Reverting Processes

The independence of relative price changes from the current price level is often criticized when commodity prices are modeled as GBM processes (e.g., Schwartz [Sch97]). It is pointed out that high commodity prices will lead to an increase in supply, as it might become profitable for higher cost producers to enter the market. The higher level of supply will have a negative effect on prices. On the other side, if prices are low, some producers might no longer be profitable and exit the market. As a consequence, supply decreases and prices tend to rise.

Mean-reverting processes are able to consider such effects. The simplest mean-reverting process, which is also known as *Ornstein–Uhlenbeck process*, follows the equation

$$dx_t = \eta(\bar{x} - x_t)dt + \sigma dz_t,$$

with $\eta > 0$ and $\sigma > 0$. Thereby, η denotes the reversion speed, $\bar{x}$ corresponds to the "normal" level of x. For commodity prices, $\bar{x}$ is the equilibrium price to which the prices are assumed to revert. As before, dz_t are the increments of a Wiener process defined in equation (1.3). The expected value of x based on the current value x_0 is

$$E(x_t) = \bar{x} + (x_0 - \bar{x})e^{-\eta t}, \tag{1.5}$$

and the variance of $(x_t - \bar{x})$ is

$$Var(x_t - \bar{x}) = \frac{\sigma^2}{2\eta}(1 - e^{-2\eta t}). \tag{1.6}$$

From equations (1.5) and (1.6) it can be seen that for large t, the expected value converges to $\bar{x}$, while the variance converges to $\sigma^2/(2\eta)$. The convergence of the expected value to the equilibrium is an advantage of a mean-reverting process compared to a GBM when used to model commodity prices. The boundedness of the variance for large t is somehow problematic as it implies that forecasts for the period $t + n$ (with $n > 0$) can be made with the same accuracy than forecasts for the period t.

Another drawback of a mean-reverting process used for commodity prices is the possibility of negative prices. For most commodities, negative prices do not exist. To preserve non-negative prices, the logarithm of the price may be used instead of the price itself. This leads to the following equation (e.g., Pilipovic [Pil98]):

$$d\ln(x_t) = \eta(\ln(\bar{x}) - \ln(x_t))dt + \sigma dz_t. \tag{1.7}$$

Note that the value $\bar{x}$ to which prices revert may be time-dependent. In such a case, we write $\bar{x}_t$. We can use $\bar{x}_t$ to model a trended mean-reverting process with fixed annual growth rates (formulated in the logarithm of the price):

$$d\ln(x_t) = \eta(\ln(\bar{x}_t) - \ln(x_t))dt + \sigma dz_t,$$

with

$$\bar{x}_t = x_0 \cdot (1 + r)^t,$$

where x_0 is the initial price and r the annual growth rate.

1.4.2 Modeling Fuel Prices

In this subsection, we briefly describe different approaches to model fuel prices. Thereby, we focus on two popular approaches, the GBM and the mean-reverting process. We conclude this subsection with some critical remarks about the quality of fuel price forecasts and the implications for the power generation expansion problem.

1.4.2.1   Literature Review

Early works in the field of fuel price modeling are often based on GBMs. For example, Gibson and Schwartz [GS90] describe the spot price of oil as a GBM. The choice of the GBM can be explained with the popularity of the famous work of Black and Scholes [BS73], who developed a model to price options based on the assumption that the price follows a GBM.

However, in the nineteen nineties, it was questioned whether the GBM assumption is pertinent. Bessembinder et al. [Bes+95] studied for several commodities whether investors anticipate mean-reversion in spot asset prices. They found mean-reversion in all examined markets. For crude oil, they reported that 44% of a typical price shock is expected to be reversed over the subsequent eight months. Schwartz [Sch97] formulates a commodity price model in which the logarithm of the spot price is assumed to follow a mean-reverting process of the Ornstein–Uhlenbeck type. Besides this one-factor model, Schwartz also formulates a two-factor and a three-factor model. The advantage of these multi-factor models is their ability to consider additional parameters influencing commodity prices. A main difficulty of such multi-factor models is that the factors that are used in these models are often not directly observable.

When fuel prices are used as an input parameter for fundamental electricity market models, the approaches the most commonly used to simulate fuel prices are GBMs (e.g., Aronne et al. [Aro+08], Botterud [Bot03]) or mean-reverting processes (e.g., Hundt and Sun [HS09]). For this reason, we focus in the following on a comparison of these two processes.

1.4.2.2   Comparison of GBM and Mean-Reverting Models

Pindyck [Pin99] studied the long-run evolution of oil, coal and natural gas prices using data of up to 127 years. Based on this data, Pindyck addressed the question whether these fuel prices should be modeled as a GBM or as a mean-reverting process. Unit root tests and variance ratio tests suggest that the logarithms of the prices are mean-reverting. However, the rate of mean-reversion is slow. Furthermore, the trends to which the prices are reverting fluctuate over time.

Even though the results indicate that energy prices follow rather a mean-reversion process than a GBM, Pindyck examined the effects of investment decisions if a GBM process is used instead. He mentions that if the rate of mean-reversion is slow, the dependence of the optimal decision on the equilibrium price is low.

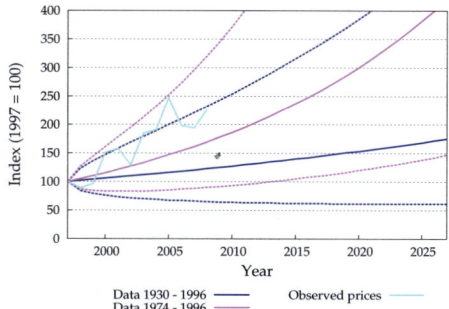

Figure 1.7: Expected gas price development and 95% confidence interval for gas prices modeled as a GBM. Drift and standard deviation estimations based on different periods. Data taken from [Pin99]. Observed gas prices taken from the US Energy Information Administration.

However, this holds only if the implied volatility of energy prices is relatively constant, as assumed by the GBM. As both conditions are fulfilled, Pindyck concludes that the GBM assumption is unlikely to lead to large errors in the optimal investment rule.

Considering the data Pindyck used for his analysis, another difficulty of predicting fuel prices becomes obvious. The annual average price changes vary significantly depending on the considered time period. While the average annual gas price increased by 1.88% in the years between 1930 and 1996, considering just the years of 1970–1996, average annual increases of 4.9% are observed. The standard deviation also increased from 0.1079 to 0.1459.

Figure 1.7 shows the expected gas prices and a 95% confidence interval for GBMs with drift and standard deviation estimated based on the two before mentioned periods. In addition, we added the observed gas prices until 2008.[14]

It can be seen that for several years, the observed prices are outside the 95% confidence interval for the GBM based on the mean and standard deviation estimations of the years between 1930 and 1996. For the GBM based on the data between 1970 and 1996, observed prices remained in the 95% confidence interval. However, this confidence interval gets quite wide.

---

[14]We took the price of gas used by electricity generators from the US Energy Information Administration (available at http://tonto.eia.doe.gov/dnav/ng/ng_pri_sum_dcu_nus_a.htm, last time accessed on December 02, 2009) and deflated the prices using the Producer Price Index for all commodities (available at http://www.bls.gov/ppi/, last time accessed on December 02, 2009) as Pindyck did it.

While the majority of researchers assume fuel prices to be mean-reverting, this is not undisputed. Geman [Gem07] examined the development of crude oil and natural gas prices during 15 years. She identified a mean-reversion pattern for crude oil prices in the years between 1994 and 2000. In 2000, the pattern changed into a random walk. For natural gas prices, a mean-reversion pattern is found until 1999. It also changed into a random walk in 2000.

1.4.2.3   Some Critical Remarks on Fuel Price Forecasts

Even though a wide variety of fuel price forecasting models exist, the quality of forecasts derived from the models is often questioned. Manera et al. [Man+07] present a survey of different models used to forecast oil prices. Manera et al. note that there is no consensus about the appropriate forecasting model. Findings vary across models, time periods and data frequencies. The authors conclude that the best performing econometric model for oil forecasts is still to appear in literature. A similar conclusion is made by Fattouh [Fat07]. The author reviewed the three main approaches used for analyzing oil prices: non-structural models, the informal approach and the demand–supply framework. Fattouh concludes that each of these approaches suffers from major limitations, especially when used to make predictions.

While those two papers surveyed oil price models, the results are probably also applicable to gas price models. At least in Germany this is the case, as the gas price is directly linked to the oil price. Summarizing the results of the literature on fuel price forecasting, two things should be retained. First, a perfect forecast model does not exist. Second, even if the appropriate stochastic process for a fuel price model could be identified, the forecasts might still differ depending on the historical data used to estimate the parameters of the process. In our opinion, these findings emphasize the importance to consider a wide variety of different fuel price scenarios in a power plant investment model.

1.4.3   Modeling Carbon Prices

Besides fuel prices, carbon prices have a strong impact on power generation costs. Similar to fuel prices, carbon prices can change the profitability of the different generation technologies. Compared to fuel prices, modeling carbon prices is even more difficult. In the following, we briefly discuss difficulties of carbon price modeling, before we give an overview of some models used in literature.

1.4.3.1 Difficulties of Carbon Price Forecasting

The additional difficulties of carbon price modeling compared to fuel price modeling are due to the short history of the EU ETS and the strong influence of uncertain future legislation.

- **Short history:** The EU ETS was started in 2005. As a consequence, the amount of historical data that can be used to develop and calibrate price forecast models is very limited. Furthermore, the first trading period from 2005–2007 can be seen as a test period. A dramatic price drop could be observed in 2006, as it became obvious that member states had over allocated allowances to emission intensive firms. It is questionable whether such a price drop is a normal market behavior which will also occur in the future, or whether it was mainly caused by immature markets.

- **Different phases:** Another problem is the division of the EU ETS into several trading periods. After phase I, the second trading period has started in 2008 and ends in 2012. As the European Union Allowances (EUAs)[15] could only limitedly be transferred between these two periods, these periods must be treated as separate markets. As a consequence, different carbon price models may be pertinent for these periods. A third trading period starts in 2013 and ends in 2020. As banking between the second and the third period is allowed, data from the second period may be used for price models considering the third period.

- **Legal framework and fundamentals:** The main price determinants for CO_2 are (i) policy and regulatory issues and (ii) market fundamentals that directly concern the production of CO_2 (Benz and Trück [BT09]). Both of them are highly uncertain, especially in the long-term. For example, the National Allocation Plans for the second trading period, which determine the amount of allowed emissions, have been finalized in January 2007. Today, only the amount of allowed emissions until 2012 is fixed. It is very difficult to predict EUA prices without knowing future regularity conditions. However, for power plant investments, long-term price forecasts are required. To the extend of the author's knowledge, there are no carbon price forecast models covering a time-horizon of several years.

---

[15]An EUA allows the emission of one tonne CO_2.

1.4.3.2   Literature Review

In the following, we give an overview of some approaches to model carbon prices. As the emission trading system is relatively new, only a few models for CO_2 price simulation exist. Most of them consider spot prices. Nevertheless, we present these results, as they describe some typical characteristics of CO_2 prices.

Paolella and Taschini [PT06] analyze the European Union CO_2 emission allowance spot prices. They discuss two different approaches to model carbon prices: an approach based on fundamentals of CO_2 and an approach based on future-spot parity of CO_2.

Following the first approach, Paolella and Taschini examine whether fuel prices can be used as a proxy for CO_2 prices. The authors report that the fuel price level does not fully explain the CO_2 price level in 2005. Concerning the second approach, Paolella and Taschini argue that due to the uncertainties related to emission allowances (e.g., political uncertainty), price scenario analysis based on the existence of a well-behaved future-spot parity is not pertinent. As an alternative, Paolella and Taschini propose different GARCH[16] models.

Benz and Trück [BT09] examine the spot price dynamics of emission allowances in the EU ETS. They propose two different models, an AR-GARCH model and a Markov regime-switching model. The best in-sample fit is reported for a regime-switching model. While the GARCH model is only slightly worse, models with constant volatility like a Gaussian distribution perform clearly worse. Benz and Trück explain the better performance of the models with conditional variance by the dependence of the carbon prices on regularity conditions and fundamentals.

Spangardt and Meyer [SM05] use a trended mean-reverting process to model carbon spot prices. Thereby, the authors assume an annual growth rate of carbon prices of 3%.

Dannenberg and Ehrenfeld [DE08] assume that carbon prices fluctuate around the expected marginal abatement costs. Hence, Dannenberg and Ehrenfeld propose to model carbon spot prices with a mean-reverting process. To consider the effects of new informations that can significantly change the expectations of marginal abatement costs, Dannenberg and Ehrenfeld introduce jumps in the level to which prices revert.

Daskalakis et al. [Das+09] analyze spot and future prices of three different markets: NordPool, PowerNext and the European Climate Exchange (ECX). The anal-

---

[16]Generalized autoregressive conditional heteroscedasticity (Bollerslev [Bol86])

ysis of spot prices suggests that spot prices are likely to be characterized by jumps and non-stationarity. A test amongst six different stochastic processes shows best results for a GBM with jumps.

In addition to spot prices, Daskalakis et al. examine two different types of futures: intra-period futures and inter-period futures. The maturity of intra-period futures is within the same emission allowances trading phase in which they are first traded. The maturity of inter-period futures is not in the same phase in which they are first traded.

While intra-period futures evolved closely with spot prices, there were significant deviations for inter-period futures. Inter-period future prices are reported to be higher and less volatile than those of intra-period futures. A possible explanation for that observation is the prohibition of banking of emission allowances between phase I and phase II of the EU ETS. For inter-period futures, the authors propose a two-factor equilibrium model based on a jump-diffusion spot price process and a mean-reverting stochastic convenience yield.

Kanamura [Kan09] studies some properties of EUA future prices traded on the ECX. The author reports that EUA future prices cannot be shown to be mean-reverting or exhibiting seasonality.

Seifert et al. [Sei+08] present a stochastic equilibrium model for CO_2 spot prices. Seifert et al. report that CO_2 spot prices do not show mean-reversion, nor do they have to follow any seasonality. The volatility of the prices is time- and price-dependent, with increasing volatility when the end of the trading period is approaching. The authors further remark that spot prices are bounded by the penalty cost plus the cost of having to deliver any lacking allowances.

A consequence of the relatively short existence of the EU ETS is the lack of established price forecasting models, as it can be seen from the brief literature summary. So far, no commonly accepted model exists for long-term price forecasts required for power generation expansion models. Therefore, most power generation expansion models use either deterministic carbon prices (e.g., Sun et al. [Sun+08], Swider and Weber [SW07]) or consider different predefined scenarios within the optimization (e.g., Jensen and Meibom [JM08], Klein et al. [Kle+08]).

1.4.4 Modeling Electricity Prices as Stochastic Processes

Electricity prices can either be modeled as a stochastic process, or they can be determined by a fundamental electricity market model. In this subsection, we discuss the first approach, the latter one is presented in detail in Chapter 2.

There are two different approaches for modeling electricity prices as stochastic processes. The first one are financial models, which focus on the stochasticity, and do not (or only marginally) consider the influence of deterministic effects. In contrast, econometric models are based on the assumption that price changes are caused by a superposition of deterministic and stochastic effects. However, the distinction is not always unambiguous. Most models presented in this subsection consider somehow the influence of deterministic effects.

In the following, we first discuss the advantages and disadvantages of electricity prices modeled as a stochastic process compared to fundamental electricity market models. Then we give a brief literature overview.

1.4.4.1   Strengths and Weaknesses

Financial and econometric models are best suited to forecast electricity prices in the short-term (see Schmutz and Elkuch [SE04]). Most of these models focus on a good approximation of the electricity spot price characteristics described in Section 1.1.3. Compared to fundamental electricity market models, financial and econometric models have the following advantages:

- In general, financial and econometric models models capture the statistical properties of electricity price distribution better than fundamental models.

- There is enough data available to calibrate these models.

- Simulations based on a financial or an econometric model are very fast compared to fundamental electricity market models.

- Financial and econometric models do not depend on the type of competition in the considered market. The only assumption is that the degree of competition does not change. For short-term periods, this condition is fulfilled.

However, financial and econometric models have two disadvantages, which make them unsuitable for long-term investment models in our opinion:

- Financial and econometric models rely on a relatively stable market structure. As the models are calibrated using historical data, price forecasts may no longer be accurate if market fundamentals change. For example, such models cannot predict the effects of increasing wind power penetration, if these effects are not yet observable on the markets.

- When such models are used for power generation expansion models, it is in general implicitly assumed that the construction of new units has no effect on observed prices. At least if investments in several units are allowed, this assumption is not pertinent. While econometric models might somehow incorporate the available capacity, fundamental electricity market models are better suited for such a case.

1.4.4.2 Literature Review

In this paragraph, a brief overview of different financial and econometric electricity price models is given. As it can be seen from the wide variety of different approaches, there is no consensus about the best method.

Deng [Den99] proposes three different mean-reverting jump-diffusion processes for modeling electricity spot price. Deng includes in these models an additional process, which may be correlated with the electricity prices. This other process can be the spot fuel prices or the electricity demand. To account for short periods of high electricity prices, which may be caused by forced plant outages, Deng extends the basic model to a Markov regime-switching model. This extended model allows to switch between normal states and high price states.

Lucia and Schwartz [LS02] formulate a pricing model for spot and derivative prices based on observations from NordPool, the Scandinavian power exchange. For the spot market, they use a simple one factor model consisting of a deterministic function depending on the time and a stochastic Ornstein–Uhlenbeck process. The deterministic function can be used to represent seasonality in the model. The one-factor model is extended by a correlated second stochastic process modeled as an arithmetic Brownian motion. The authors stress that the inclusion of the deterministic factor accounting for regularities in the behavior of electricity prices is important. At the same time, Lucia and Schwartz point out that it is difficult to estimate this seasonal pattern, as it requires a time series of several years.

Barlow [Bar02] uses a simple supply/demand model[17] to determine electricity spot prices. The electricity demand is modeled as an Ornstein–Uhlenbeck process. The resulting model for electricity spot prices is a diffusion model including price spikes.

---

[17]This model could also be classified as a fundamental model, as it considers the supply curve as described in Section 2.2. On the other side, the supply is modeled as a stochastic process, and the model is discussed from the point of view of statistical models.

Escribano et al. [Esc+02] present a multi-factor model for average daily spot prices. The model considers seasonality, mean-reversion, stochastic volatility and jumps. To allow jumps in their model, Escribano et al. use an explicit jump term.

Burger et al. [Bur+04] develop a model considering seasonality, mean-reversion, price spikes, price-dependent volatilities and long-term non-stationarity. Their model is a three-factor model consisting of a stochastic load process, a short-term process and a long-term process. While the stochastic load process combined with the deterministic supply function represents the part of the price explained by fundamental data, the short-term and long-term process are used to account for price determinants like psychological facts or the behavior of speculators. The load process and the short-term process are modeled as a seasonal ARIMA[18] process, while the long-term process is modeled as a random walk with drift.

Geman and Roncoroni [GR06] introduce a "jump-reversion" component in their spot price model to account for price jumps. They define two different states based on a threshold value. If the current price is below the threshold, only positive jumps can occur. If the price is above, only negative jumps are possible. The intensity of the jumps may be time-dependent.

Weron and Misiorek [WM08] compare 12 time series methods for short-term spot price forecasting. The considered methods include autoregressive models, regime switching models and mean-reverting jump diffusions. The performance of the models is tested based on time-series from the Californian market and from the Nordic market. Weron and Misiorek find that for point forecasts, models including the system load as exogenous variable perform better than models without, at least for the Californian market. Using the air temperature as exogenous variable does not necessarily improve model quality.

As this literature survey shows, there are many different approaches and stochastic processes used to model electricity prices. A common point of most models is the inclusion of fundamentals to improve price forecasts.

1.4.5 Modeling Electricity Prices with Fundamental Models

Fundamental electricity market models calculate the cost-minimal electricity production for the considered market. They can be used to determine the electricity production of each unit as well as electricity prices. The main advantage of fundamental electricity market models compared to simulation models is the more

---

[18]AutoRegressive Integrated Moving Average, see, e.g., Box et al. [Box+08]

appropriate consideration of the relationship between fuel prices, electricity demand and electricity supply on the one side and electricity production and electricity prices on the other side.

Depending on the type of fundamental market model, they can also consider short-term uncertainties, e.g., the feed-in from wind power, as well as technical restrictions of power plants like minimum up-times or reduced part load efficiency. For these reasons, we argue that fundamental electricity market models are best suited to model electricity prices for the valuation of power plant investments.

However, fundamental models face several difficulties, which should be mentioned in the following:

- Most fundamental electricity market models (including the ones we use in this thesis) are based on the assumption of a competitive market. As it was discussed in Section 1.2.2, this assumption is still questionable nowadays. However, the planned extension of cross-border transmission capacity will increase competition on national markets, and may result in a common European electricity market. Under such circumstances, the assumption of perfect competition will be more appropriate.

- In order to be able to make long-term price forecasts, assumptions about future market conditions like fuel prices or the existing power plant fleet must be made. As we saw in Subsections 1.4.2 and 1.4.3, this is quite difficult.

- Fundamental electricity market models cause a high computational burden. Especially LP or MILP models have significantly longer running times than simulation models.

- A further problem of fundamental models is data availability. While the data required for financial and econometric models is in general publicly available (e.g., historical electricity prices), the exact technical parameters of power plants are often not publicly known.

Despite these disadvantages, we belief for the before mentioned reasons that fundamental electricity market models are the appropriate way to determine electricity prices for power plant valuation methods. We describe different types of fundamental electricity market models in detail in Chapter 2. In the following, we briefly outline three different kinds of fundamental electricity market models, because they are used afterwards in the literature review in Section 1.6.

- **Load–duration curve (LDC):** Fundamental models based on the LDC are the simplest type of fundamental models. The LDC for a whole year is used to determine the corresponding electricity production and prices. LDC models do not consider any technical constraints.

- **Supply–demand curves (SDC):** These models use several different SDCs to determine electricity prices. As the models based on the LDC, these models do not consider technical constraints. However, models based on the SDC can consider a price elastic demand. As the application of a SDC model delivers only results for one moment in time, more computations are required compared to LDC models.

- **LP/MILP models:** Describing the power generation as an LP or an MILP problem makes it possible to model technical constraints like minimum operation or minimum shut-down restrictions. Furthermore, reduced part-load efficiency and short-term uncertainties like the demand or the feed-in from wind power can be considered. The disadvantage of such models is the high computational burden.

1.5   Valuation of Power Plant Investments

The uncertainties discussed in Section 1.3 have a strong influence on the profitability of power plants investments. Hence, these uncertainties have to be considered by power plant valuation methods.

In this section, we present two different approaches to optimize power plant investments. The first approach is based on the *Levelized Costs of Electricity Production* (LCEP). This method calculates the expected costs of electricity production for different technologies and selects the technology with the lowest cost. The LCEP approach was mainly used in a regulated electricity market. It does neither consider the timing of investments, nor does it consider the before mentioned uncertainties in an appropriate way.

Owing to these shortcomings of the LCEP approach, the real options approach, a method developed for investments under uncertainty, became popular to evaluate power plant investments in a deregulated market. Other than the LCEP approach, it does explicitly address the timing of investments.

In the following, we first describe the LCEP approach, as it still seems to be applied in practice (see, e.g., Konstantin [Kon08]). Then, we introduce the real options approach, which we use for our investment models.

1.5.1 Levelized Costs of Electricity Production

The LCEP is based on the *Net Present Value* (NPV) approach, a standard method to evaluate investment opportunities (see, e.g., Brealey et al. [Bre+07]). The NPV is the sum of a series of cash-flows discounted to its present value. As discount factor, the firm's weighted average cost of capital may be used. Formally defined, the NPV is

$$\text{NPV} = \sum_{t=0}^{T} \frac{R_t}{(1+r)^t},$$

with R_t equal to the returns in year t, r the discount rate and T the planning horizon. To emphasize the initial investment I_0 and a possible salvage value L at the end of the planning horizon, it is often written as

$$\text{NPV} = -I_0 + \sum_{t=1}^{T} \frac{R_t}{(1+r)^t} + \frac{L}{(1+r)^T}.$$

Based on the NPV criterion, an investment is carried out if the NPV is positive. If one of several investment alternatives has to be chosen, the one with the highest NPV is realized.

Even though the NPV approach has some pitfalls (e.g., choosing an appropriate risk-adjusted discount rate), it is widely used in practice. A survey of 392 chief financial officers of large and small firms showed that 74.9% of them use always or almost always the NPV method for project valuation (Graham and Harvey [GH01]).

The LCEP correspond to the average costs of electricity production in €/MWh considering all discounted costs that occur during the system lifetime. They are calculated in such a way that the NPV of a possible power plant investment is zero. The LCEP are defined as (see, e.g., OECD and IEA [OI05][19])

$$\text{LCEP} = \frac{\sum_{t=0}^{n} \frac{I_t + M_t + F_t + E_t}{(1+r)^t}}{\sum_{t=0}^{n} \frac{P_t}{(1+r)^t}},$$

---

[19]The emission allowance expenditures are not yet considered by the definition given in [OI05], probably due to the date of the publication. As emission allowance expenditures are costs related to the electricity production, we follow [Kon08] and include them in the LCEP.

with I_t = Investment expenditures in year t [€],

M_t = Operation and maintenance expenditures in year t [€],

F_t = Fuel expenditures in year t [€],

E_t = Emission allowance expenditures in year t [€],

P_t = Electricity production in year t [MWh],

r = Discount rate,

n = System lifetime.

To evaluate different investment alternatives, the LCEP are calculated for each investment alternative. Then, the project with the lowest LCEP is chosen.

An example of an LCEP calculation is depicted in Figure 1.8(a). It shows the LCEP for different units based on the assumptions of an 85% average load factor (corresponding to 7,446 full load hours per year), a discount rate of 10% and carbon prices of 20 €/t. It can be seen that the different cost blocks for a coal unit are about the same size, while the costs of a CCGT unit are mainly influenced by the fuel costs. The most important costs for nuclear units are the investment costs.

The LCEP concept was a popular method for project valuation in a regulated market environment. However, it has several severe shortcomings, which are described in the following.

- **No appropriate consideration of uncertainty:** The LCEP calculation is based on expected values, ignoring any kind of uncertainty. To get an idea what happens when the value of one of the parameters deviates from the expected values, a sensitivity analysis is usually performed. For this purpose, the LCEP are calculated as a function of a chosen input parameter. Besides this chosen parameter, all other parameters remain fixed.

    An example of such a sensitivity analysis is shown in Figure 1.8(b). It shows the LCEP for lignite plants, coal plants, CCGT plants and gas turbine plants depending on the number of full load hours. It can be seen that above about 3,500 full load hours, lignite plants are the cheapest option. If the number of full load hours is less, CCGT units are cheaper. For this example, the investment in lignite plants seems to be a robust decision, as the threshold value changing the ranking of the technology is far away from the assumed number of full load hours for lignite units, which is 7,500 [Kon08]. Even though a sensitivity analysis gives meaningful insights, it also suffers from some limitations.

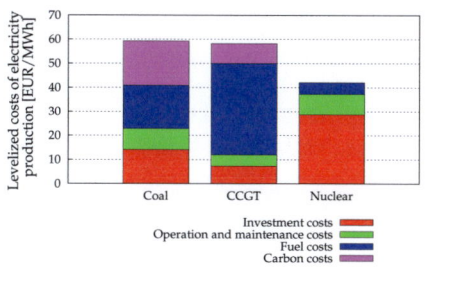 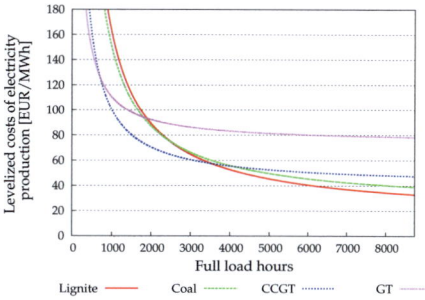

(a) LCEP of coal, CCGT and nuclear units. Source: OECD and IEA [OI05].

(b) Sensitivity analysis for the LCEP of lignite, coal, combined cycle gas-turbine and gas-turbine units. Source: Konstantin [Kon08].

Figure 1.8: Levelized Costs of Electricity Production

First of all, the sensitivity analysis varies only the value of one parameter and assumes the values of all other parameters to be fixed. Second, while it varies the value of one parameter, it is assumed that the changed values remain constant throughout the whole planning horizon. For example, a sensitivity analysis that considers the impact of gas prices still assumes that the alternative gas prices are constant during the whole optimization horizon. It compares scenarios with gas prices of 10 €/MWh and 15 €/MWh for example, but it does not consider scenarios where gas prices change from year to year, e.g., from 10 €/MWh in the current year to 15 €/MWh the next year.

Compared to reality, this is a very simplistic consideration of uncertainty. Especially fuel prices and carbon prices exhibit a high volatility. As the profitability of the different technologies strongly depends these prices, fuel and carbon price uncertainty must be considered adequately by a power plant valuation method.

- **Assumptions about electricity production:** To calculate the LCEP, assumptions about the expected electricity production are made. While this approach was pertinent in a regulated market, it is no longer applicable in a deregulated market. In a competitive electricity market, a unit produces only if its marginal costs are lower than or equal to the electricity prices.

- **No consideration of electricity prices:** Another drawback of the LCEP approach is its focus solely on the cost side. Blindly applying the LCEP method ignores that peak load units operate mainly during periods with high prices. Hence, they may be profitable in spite of their relatively high costs.

- **No consideration of operational flexibility:** Due to the increasing amount of electricity produced by intermittent sources like wind power, operational flexibility becomes more and more important. Power plants with short start-up times, which are able to react on wind power fluctuations, will gain market shares from inflexible units. The LCEP approach does not consider this value of operational flexibility.

- **No evaluation of the profitability of investments:** The LCEP method does not explicitly address the question whether an investment should be carried out or not. It only determines the cheapest technology based on the assumption that an investment is required. In a regulated market, the generation companies were obliged to guarantee the security of supply. Based on the existing reserve capacity and the expected demand growth, they could decide whether new investments were required or not. In a deregulated market, generation companies will only invest if an investment is profitable. To be able to answer this question using the LCEP approach, one may compare the current electricity prices with the calculated LCEP. However, such a comparison does not consider the uncertainty related to future electricity prices.

- **No consideration of timing:** Additionally to the question whether to invest or not, also the question when to invest must be answered. The NPV approach (and also the LCEP approach) are based on the implicit assumption that the investment is either reversible or that the investment opportunity is a "now or never" opportunity. Reversibility of investments means that the investments may be undone in the future and the costs can be recovered somehow. The "now or never" opportunity implies that the investment is either realized immediately, or never. However, these two conditions do not hold for power plant investments. Once the investments are carried out, the investments can hardly be undone. Investment costs must be considered as sunk costs which cannot be recovered. In most cases a company has the possibility to postpone power plant investments. Hence, methods based on the NPV are not appropriate for power plant investments.

Due to these shortcomings of the NPV method, an alternative method, the *real options approach*, became popular for power plant valuation.

1.5.2 Real Options Approach

The real options approach is based on the observation that the possibility to delay an irreversible investment has some additional value which must be included in a project valuation (see, e.g., McDonald and Siegel [MS86]). The term "real options" was introduced by Myers [Mye77]. Myers noticed that a firm is composed of two asset types: real assets, which have market values independent of the firm's investment strategy, and real options, which are opportunities to purchase real assets on possibly favorable terms.

The concept of real options is closely related to financial options. Having the opportunity to invest is analogous to holding a financial call option—the holder has the right, but not the obligation to buy an asset in the future. Carrying out the investment corresponds to exercising the option. Investing "kills" the option. Contrary to the NPV analysis, it is not sufficient for an investment to have a positive NPV to be carried out, but the NPV of the immediate investment must exceed the option value.

In the following, we use a simple example to demonstrate the main idea of the real options approach. Then, we discuss the influence of different problem parameters like the degree of uncertainty on investment decisions. In the second part of this subsection, we give a brief introduction to SDP and demonstrate how SDP can be used to model power plant investment problems based on the real options approach.

1.5.2.1 Introduction

The value of the option to postpone an investment is closely related to the uncertainty of the project's future cash flows. This can be illustrated with a simple example.

Example. *Assume that an electricity generation company has the possibility to invest either in a new coal plant or in a CCGT plant. Assume further that the future contribution margins strongly depend on future carbon prices, which themselves depend on a political decision concerning the amount of allowed emissions. However, this decision is taken in one year, so its outcome is not yet known. There are two possible carbon price scenarios with equal probability. Postponing the investment results in lost discounted cash flows (DCFs) for both alternatives. The assumed payoffs for the two investment alternatives are summarized in Table 1.3.*

	Coal unit	CCGT unit
Investment costs	1,000	500
DCFs High carbon prices	1,050	650
Low carbon prices	1,200	500
Lost DCFs due to postponing the decision	50	30

Table 1.3: Data used for a simple example to demonstrate the value of postponing an investment.

Applying the NPV method results in a NPV of 125 ($-1,000 + 0.5 \cdot 1,200 + 0.5 \cdot 1,050$) for the coal unit and 75 ($-500 + 0.5 \cdot 650 + 0.5 \cdot 500$) for the CCGT unit. As both NPVs are positive and the NPV of the coal unit is greater than the NPV of the CCGT unit, one would choose to invest in the coal unit (remember that only one investment is allowed).

Postponing the investment by one year reveals additional information. In case of low carbon prices, the NPV of the coal unit is 150 ($-1,000 + 1,200 - 50$), while the NPV of the CCGT unit is -30 ($-500 + 500 - 30$). If the high carbon price scenario occurs, the NPV of the coal unit is 0 and the NPV of the CCGT plant is 120. If the investment is postponed, then the optimal decision depends on the realized scenario. In case of low carbon prices, the coal unit is preferable. In the case of high carbon prices, the CCGT unit is the better choice. The expected NPV of postponing the investment is 135 ($0.5 \cdot 150 + 0.5 \cdot 120$). This value corresponds to the option value. As the NPV for both immediate investment alternatives is lower than the option value, the optimal decision is to postpone the investment.

One should note that the fact that the optimal decision depends on the realization of the uncertain carbon prices is not a sufficient condition for postponing the investment. Assume that the lost DCFs for postponing the investment are now 70 for the coal unit and 50 for the CCGT unit. If the investment is postponed, it is still preferable to build the coal unit under the low carbon price scenario and the CCGT unit under the high carbon price scenario. However, the option value drops to 115 due to the increased lost DCFs. Now, the NPV of an immediate investment is higher than the option value. Assuming a risk-neutral investor, the optimal decision is to invest immediately.

Real options have some common characteristics, which are described for example by Dixit and Pindyck [DP94]. We summarize them in the following, as they allow to gain interesting insights on the influence of some parameters on investment decisions in general, which are also important for our investment model.

- **Changing the cost of the investment:** The option value associated to a possible investment increases with the sunk costs of the investments. As a consequence, the incentive to delay capital-intensive investments like coal units is higher than the one for cheaper units like CCGT plants. Such a behavior is for example observed by Olsina [Ols05].

- **Degree of uncertainty:** Increasing the degree of uncertainty leads to a higher option value. On the one hand, the downside risk of the option is zero. If unfavorable future scenarios occur resulting in a negative NPV, the option will not be exercised (no investments will be done). On the other hand, the upside payoff of the option increases. As increasing the volatility of uncertain payoffs without changing the expected values has no influence on the NPV of an immediate investment, an increase of volatility delays investments.

- **Influence of hedging:** The ability to hedge risk with contracts on forwards or future markets does not change the option value. This fact can be explained by the observation that on efficient markets, risks are fairly priced. Any decrease of risk correlates with a decrease in return. As a consequence, the possibility to hedge does not need to be considered in an investment model based on the real options approach.

- **Effects of interest rates on investments:** A reduction of interest rates leads to a higher NPV. Hence, it increases the incentive for investments. At the same time, it makes future cash flows more important relative to present ones. As a consequence, the uncertainty increases, and with it the option value. These two opposite effects explain why interest rates seem to have only little effect on investments. Even though interest rates are still an important parameter, they have less impact when the real options approach is applied to valuate power plant investments compared to the case when the LCEP approach is used.

1.5.2.2 Using Stochastic Dynamic Programming to Model Real Options

To solve real options problems, either SDP or the contingent claims approach can be used [DP94]. The concept of contingent claims analysis is closely related to financial option pricing. To be able to apply the contingent claims analysis, a portfolio of traded assets which exactly replicates the pattern of returns of the investment project is required. It is at least questionable whether this condition is met in the case of electricity power plants.[20] For this reason, we opt for the SDP approach in this thesis. In the following, we briefly describe the main ideas of SDP.

Dynamic programming (DP) goes back to Bellman [Bel57]. The main idea of DP is to break a complicated problem down to smaller subproblems which can be solved recursively. A discrete-time dynamic optimization problem consists of state variables x_t and decision variables a_t. The set of states at time t is denoted by $\mathcal{X}_t$, the set of possible decision depending on the current state x_t is denoted by $\Gamma(x_t)$. Based on the transition function $T_{t+1}(x_t, a_t)$, a new state x_{t+1} is reached when decision a_t is applied while being in state x_t. In case of a stochastic problem, to which we refer as SDP, we write the transition function as $T_{t+1}(x_t, a_t, w_{t+1})$, where w_{t+1} denotes the state of the stochastic process. Associated with each action a_t which is taken at state x_t is an immediate payoff $F(x_t, a_t)$. The optimal value, written as a function of the state is the value function $V_t(x_t)$.

The DP approach calculates for each state the profit maximizing decision. In other words, DP calculates the optimal policy, where the policy determines for every state which action is taken. Using the Bellman equation, the problem can be recursively defined as:

$$V_t(x_t) = \max_{a_t} \left(F(x_t, a_t) + \beta \cdot \sum_{x' \in \mathcal{X}_t} \left(P(x_{t+1} = x' \mid x_t, a_t) \cdot V_{t+1}(x_{t+1}) \right) \right)$$
$$\text{subject to} \quad a_t \in \Gamma(x_t)$$
$$x_{t+1} = T(x_t, a_t, w_{t+1}).$$

Thereby, $P(x_{t+1} = x' \mid x_t, a_t)$ denotes the probability that x' is reached when decision a_t is taken while being in state x_t, and β denotes a discount factor with $0 \leq \beta \leq 1$.

This problem can be solved using the backward induction. Let T be the number of time steps considered. We start with solving the problems of the last stage, $V_T(x_T)$ for all state variables x_T. Using this solution, we can solve $V_{T-1}(x_{T-1})$. We continue stepping backwards through time, until we solve $V_0(x_0)$.

To be able to apply DP, some preconditions must be met. First, the cost functions must be additive over time. Furthermore, the optimal decision at state x_t must be independent of the previous decisions.

---

[20]For example, derivatives are traded only for six years in advance at the EEX. Hence, it is impossible to construct today a portfolio which replicates the cash-flows obtained by a new power plant for the next 30 to 40 years.

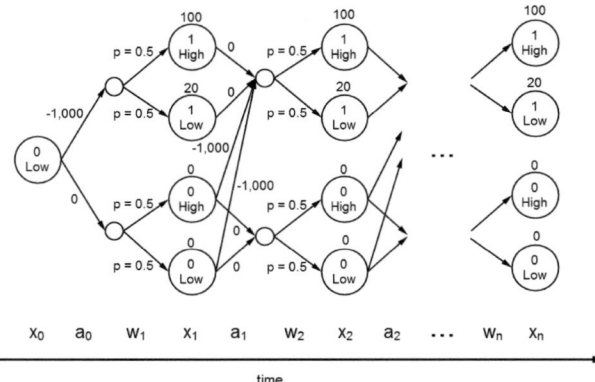

Figure 1.9: Simple example for a power plant investment problem.

If an SDP problem is considered, as it is the case for real options problems, then it is important that the stochastic process is Markovian, i.e., the transition probability depends only on the current state of the process, but not on its history. For an extensive introduction to SDP see, e.g., Bertsekas [Ber05].

Example (Power plant investments formulated as an SDP problem). *Let us briefly outline how SDP can be used to evaluate power plant investments. Assume that the investment in one coal plant should be evaluated. The investment costs are 1,000. The possible decisions are either to invest or not invest. This may be encoded the following way:*

$$a_t = \begin{cases} 1 & \text{if the investment is carried out,} \\ 0 & \text{otherwise.} \end{cases}$$

A state can be composed of several attributes, e.g.,

$$x = \begin{pmatrix} \text{Number of new units} \\ \text{Electricity price} \end{pmatrix},$$

with number of new units equal to 0 or 1 and electricity prices being low or high. Assume that there is a probability of 0.5 to switch from low electricity prices to high electricity prices and vice versa. Let the payoff in the case of high electricity prices be 100, in the case of low electricity prices 20. Figure 1.9 visualizes this problem.

The initial state x_0 is characterized by low electricity prices and no existing units. We then take our decision a_0 before the outcome of the stochastic event w_0 is revealed. Depending on our decision and the stochastic event we get to the next state x_1. Assume that we take the decision to build the new plant, and that the electricity prices during the next period

are high. The payoff $F(x_0, a_0)$ is $-1,000$, the one for $F(x_1, a_1)$ is 100. To solve this simple
problem, the backward induction approach can be used.

While the given example is a very simplistic example, modeling power plant
investments as real options using SDP generally follows the outlined approach.
However, there are several design decisions in which the existing models differ.
The most important design decisions are the choice of the investment alternatives,
the way the contribution margins for the new units are determined and the parameters for which uncertainty is considered. We discuss several alternatives in the
next section, where we give an overview of related work.

1.6   Related Work

Power generation expansion planning has attracted a lot of research in recent
years. Depending on the objective of the research, a wide variety of different models is applied. In this section, we give an overview of previous work classified by
the chosen approach.

Different classifications of decision support models in long-term electricity planning exist. Ventosa et al. [Ven+05] distinguish single-firm optimization models,
equilibrium models and simulation models.

Weber [Web05] differentiates between descriptive and prescriptive models. Descriptive models try to describe how things actually work, while prescriptive
models say how things should work. On the first sight, it may seem that a classification of a model as either descriptive or prescriptive is unambiguous. However,
as pointed out in [Web05, p. 27], also prescriptive models may contain a description of optimizing agents, and might therefore also be formulated as a descriptive
model.

Olsina [Ols05] uses basically the same classification as Weber, however he uses
the terms optimization models (instead of prescriptive) and simulation models
(instead of descriptive). The difficulty to classify some models can be seen comparing classifications of game-theoretic models. While Sensfuß [Sen07, p. 24] clas-
sifies them as simulation models, Olsina [Ols05, p. 6] treats them as optimization
models.

To avoid this difficulty, we follow the classification applied in [Ven+05], treating
game-theoretic models as a separate class of models. In the following, we describe simulation models, game-theoretic models and optimization models. As

our model is an optimization model, we review optimization models in more detail. For these models, we consider some additional classification criteria proposed by Botterud [Bot03]. These additional criteria are the representation of supply, demand and the electricity market and the representation of investment decisions.

A detailed discussion about the strengths and weaknesses of the different types of models can be found in [Sen07]. A survey comparing different approaches for power generation expansion planning from monopoly to competition is given by Kagiannas et al. [Kag+04].

1.6.1 Simulation Models

The main objective of descriptive electricity market models is to gain insights about the way the electricity market works. Using this knowledge about the functionality of the market, these models aim for predicting the future development of the market. Descriptive models do not rely on the assumption of a perfect market, but can consider market imperfections. This is an important feature, as electricity markets show different types of imperfections (see, e.g., IEA and OECD [IO03], Lemming [Lem05]). Descriptive models are based on a simulation. Two different methodologies are applied to study the long-term development of electricity markets: system dynamics and agent based simulation.

1.6.1.1 System Dynamics

The system dynamics approach goes back to Forrester [For58]. System dynamics describes the development of a complex system with stocks and flows. Feedback loops are used to model the interdependencies between the components of the system under study. A famous example of a system dynamics model is *The Limits to Growth – A Report for the Club of Rome's Project on the Predicament of Mankind* [Mea+72]. This model was built to analyze five major trends of global concern: accelerating industrialization, depletion of nonrenewable resources, a deteriorating environment, rapid population growth, and widespread malnutrition.

There are some applications of system dynamics to long term electricity market modeling. Bunn and Larsen [BL92] use a system dynamics model to test how power plant investments are influenced by various regulatory conditions, economic assumptions and the strategic behavior of the separate companies. Ford

[For01] simulates power plant investments in the Californian market. The dominant investment pattern follows a boom and bust cycle similar to the one observed for example in the real estate industry. Vogstad [Vog04] uses a system dynamics model to analyze long-term versus short-term implications of various energy policies within the context of the Nordic electricity market. Concerning power generation expansion, he also observes the boom and bust cycle. The same is true for Olsina [Ols05], who investigates the long-term development of the generation capacity. Thereby, he notices a preference for less capital-intensive technologies.

1.6.1.2 Agent Based Simulation

While the system dynamics approach assumes a centralized decision maker, the agent based simulation considers multiple independent agents. The agents may behave differently (e.g., based on their risk aversion) or they may face different restrictions. In addition, agents are able to learn. Agent based modeling of electricity markets became popular during the last years. For recent surveys see Sensfuß et al. [Sen+07] or Weidlich and Veit [WV08].

Czernohous et al. [Cze+03] present a long-term energy model considering investment decisions. Three different types of agents are modeled: suppliers, customers and a regularity authority. The electricity prices are determined by an auction, where the suppliers and customers place their bids. Two different planning layers are used. In the short-term planning layer, the suppliers decide on the plant utilization and the bids for the auction. The long-term planning layer is used to determine the investment decisions. The investment decisions are formulated as an LP profit maximization problem. Thereby, the profit calculations are based on the electricity prices observed during the last short-term period. The regularity authority decides on taxes for emissions of harmful substances.

Genoese et al. [Gen+08] study the impact of different emission allocation schemes on power plant investments in the German electricity market. The work is based on the PowerACE model, an agent based simulation of the German electricity market. The model includes a short-term spot market, a forward market, an emission allowances market and a balancing market. The prices obtained at the spot market and the forward market are used to forecast electricity prices, which are required for the investment planner. Based on these predicted prices, the investment planner calculates the net present value for the different investment alternatives. Five different emission allowance allocation schemes are tested. It is observed that the emission allowance trading leads to increasing electricity prices and that the

design of the allocation scheme has a significant impact on power plant investments.

1.6.2 Game-Theoretic Models

The focus of game-theoretic models lies on the interaction of different market players. Game-theoretic models are especially well suited for markets with imperfect
competition like oligopolies, where strategic behavior can be observed. However,
they often use a simplified description of the electricity market. Uncertainty, e.g.,
about future fuel prices, is not considered or only in a simplistic manner. Game-
theoretic models are mainly used to analyze market power and test different market designs. While the majority of the game-theoretic models focuses on short-
term questions, there are also some models considering capacity expansion.

Chuang et al. [Chu+01] study the effect of different levels of competition on capacity expansion. Ventosa et al. [Ven+02] compare capacity expansion under different assumptions about the timing of the investments. They use a cournot-based
model where the players decide simultaneously on their investments. The results are compared with a stackelberg-based model, in which a leader decides
first on its investments. Murphy and Smeers [MS05] study capacity expansion
in an oligopolistic market. They compare the case of selling electricity in long-
term contracts with the case when electricity is sold on a spot market. Pineau and
Murto [PM03] analyze capacity expansion in the Finish market under uncertain
demand growth.

1.6.3 Optimization Models

Prescriptive models, also called normative models, aim to identify the optimal decision for the considered decision maker. Most optimization models rely at least
partly on the assumption of a perfect competition. The majority of recent optimization models in the field of power generation expansion is based on the real
options approach. While these models are not able to represent strategic behavior as it can be done with a game-theoretic approach, they generally model the
electricity market and the uncertainties related to power generation expansion in
more detail.

Amongst the optimization models, there are important differences in the way the
investment decisions and the electricity market are represented. Concerning the

representation of investment decisions, Botterud [Bot03] distinguishes three perspectives: centralized decision making, decentralized decision making considering multiple decision makers and decentralized decision making considering a single decision maker. The representation of the electricity market can either be based on simulation, or on a fundamental electricity market model.

In the following, we first discuss the differences between centralized and decentralized decision making, before we present different power generation valuation approaches based on electricity price simulation. Then, we give a detailed overview of power generation expansion models based on fundamental electricity market models.

1.6.3.1 Representation of Investment Decisions

Centralized decision making can be observed in regulated electricity markets. As these markets are monopolies, there is one single decision maker who controls the whole system. From the modeling approaches presented, the system dynamics approach as well as some optimization models use a centralized decision maker.

However, in deregulated electricity markets, there are several competing actors on the market. In such an environment, a decentralized approach considering different market players is more realistic. Decentralized decision making can be further distinguished. The first possibility is to model several individual decision makers interacting through the power market. The decision makers may have individual objectives or may be restricted by individual constraints. Such an approach is taken by agent based models and also by game-theoretic models. The other possibility is to optimize the investment of a single decision maker, and represent all other decision makers as an aggregated decision maker. The decisions of this aggregated decision maker may be fixed externally or depend on the feedback from the power market. Such an approach is taken by some optimization models. We also follow this approach.

It is often stated that in a competitive market, the social welfare maximization problem leads to the same result as the profit maximization for single generation companies (e.g., Klein et al. [Kle+08], Weber and Swider [WS04]). Hence, the decentralized investment problem may be replaced with a centralized approach.

However, as discussed by Botterud [Bot03], there are several special characteristics of a power market that can distort the social welfare equilibrium. These factors are the limited price elasticity of demand, a price cap introduced by regulators, the potential exercise of market power and the lumpiness of investments. Botterud compared the outcomes of centralized and decentralized investment models, and concludes that under the centralized model, investments are higher.

From the modeling point of view, it makes a great difference whether a centralized or a decentralized decision maker is chosen. If the contribution margins for new units are determined by a fundamental market model based on an LP or an MILP formulation, then using a centralized decision maker makes it possible to integrate the investment decisions into the existing model. This is possible because the problem can be formulated as a cost minimization problem subject to demand satisfaction.

Taking the perspective of a single, decentralized decision maker, the demand satisfaction constraint no longer exists in a deregulated market. The objective of the considered decision maker is profit maximization. The profit depends on the amount of produced electricity, the costs for producing the electricity and the electricity prices. While the amount of produced electricity is a variable in the unit commitment problem and the costs can also be expressed as a variable in the unit commitment problem, the electricity prices are indirectly determined as dual variables of the electricity demand constraint.[21] This indirect determination of electricity prices makes it impossible to incorporate the investment decisions into a decentralized unit commitment model.

As a consequence, unit commitment decisions and investment decisions must be taken by two separate models. The investment problem can be formulated as an SDP problem as briefly outlined in Subsection 1.5.2.2. However, as a consequence, the unit commitment must be optimized for all states of the SDP problem. Depending on the number of considered investment alternatives and the number of stochastic scenarios, this number can be huge.

1.6.3.2 Pure Simulation-Based Models

The electricity price models presented in Section 1.4.4 can be used for power plant valuation. To determine the profits of new units, assumptions must be made about the amount of produced electricity. A common assumption is that a unit produces when the simulated electricity prices are higher than the marginal costs of the unit.

Deng [Den05] studies the option to invest in a gas plant based on stochastic gas and electricity spot prices. The model contains spikes and jumps in the spot prices.

---

[21]See Section 2.3.5.

It is shown that the presence of jumps and spikes can have a strong impact on investment decisions. While upward jumps increase the option value (and therefore delay investments), downward jumps have a contrary effect.

Instead of spot prices, Fleten and Näsäkkälä [FN03] use stochastic forward prices to evaluate the option to invest in a CCGT unit. Upper and lower bounds for investment thresholds are determined depending on the level of operational flexibility of the plant. It is shown that the operating flexibility has a significant impact on the investment thresholds.

Keppo and Lu [KL03] incorporate in their model a penalty for the investment which should reflect the price impact of the new unit on the existing portfolio. It is shown that if an investor has an exclusive option to invest, this penalty significantly increases the investment threshold. However, if other investors may also carry out the investment, then this is no longer the case.

Other than the before mentioned contributions, Tseng and Barz [TB02] incorporate operational constraints like minimum up-time and minimum down-time constraints in their power plant valuation model. The valuation is based on spot prices modeled as stochastic mean-reversion processes. A central result of the paper is the fact that ignoring operational constraints can lead to a significant overestimation of the plant value.

1.6.3.3 Models Based on Fundamental Electricity Market Models

Table 1.4 gives an overview of the models that are the most similar to our approach. It can be seen that there is no consensus which uncertainties should be considered.[22] These models and some more are described in the following paragraphs. Thereby, we distinguish between models following a centralized approach, and models opting for a decentralized approach.

Centralized Decision Making

Swider and Weber [SW07] use a stochastic fundamental electricity market model to estimate the costs of wind's intermittency. The model also incorporates investments in new capacity. The focus of the model lies on the consideration of short-term uncertainty in the unit commitment caused by stochastic feed-in of wind power. Wind power is modeled using a recombining scenario tree with three different wind power scenarios for each timestep. Transition probabilities for the

---

[22]It should be noted that the three models considering short-term wind fluctuations are based on the same underlying fundamental power market model.

	Investors perspective	Power market model	Short-term uncertainties	Long-term uncertainties
Swider and Weber [SW07]	c	LP	Wind	-
Sun et al. [Sun+08]	c	MILP	Wind	-
Klein et al. [Kle+08]	c	LDC	-	CO_2, F
Botterud et al. [Bot+05]	c/dc	SDC	-	D
Hundt and Sun [HS09]	dc	LP	Wind	F
Jensen and Meibom [JM08]	dc	LP	-	CO_2, Wind, T

Table 1.4: Overview of similar approaches to ours. Legend: c = centralized, dc = decentralized, D = demand, F = fuel prices, T = additional transmission capacity

different scenarios are derived from historical data. However, no long-term uncertainties are considered in the model. As the model is formulated from a centralized point of view, the unit commitment and the investment decisions are taken within the same model. Investments are allowed in coal units, CCGT units, gas turbine units and lignite units. Applying their model to a case study considering the German electricity market, the authors conclude that a static, deterministic model overestimates the value of intermittent energy sources. The reasons for the higher costs in the stochastic model are the increased hours of partload operation and increased start-up costs as well as a higher amount of required investments in conventional plants.

Sun et al. [Sun+08] extend this model with a mixed-integer formulation and further technical constraints like start-up time for example. In addition, combined heat and power (CHP) plants are modeled in more detail and investments in CHP units are added. The case study confirms the result from [SW07] that ignoring short-term uncertainty leads to an underestimation of the total investments required.

Klein et al. [Kle+08] optimize the power plant investments in the US market using an SDP formulation. They consider three different scenarios for fuel and carbon prices. However, only one switch from the reference scenario to either the low price or high price scenario is allowed. The electricity production is determined based on an LDC model. The problem is formulated as a cost minimization problem subject to demand satisfaction.

Contrarily to the majority of power generation expansion models, Meza et al. [Mez+07] formulate the generation expansion planning as a multi-objective deterministic optimization problem. Besides the sum of the total costs, the CO_2 emissions, the amount of imported fuel and the energy price risk are considered.

Newham [New08] uses stochastic dual dynamic programming to solve a generation expansion planning problem with stochastic demand growth. In addition to the generation expansion, also the transmission network extension is considered.

Decentralized Decision Making

Göbelt [Göb01] uses a combination of a single-stage and a multi-stage stochastic optimization model to optimize power plant investments. While electricity and heat demand are modeled using a multi-stage stochastic programming formulation, fuel prices are modeled based on the single-stage stochastic programming approach. Thereby, expected fuel prices are increased by a penalty term that reflects the uncertainty related to the prices. This penalty term is obtained by multiplying the standard deviation of the fuel prices with a factor representing the degree of risk aversion of the considered operator.

Botterud et al. [Bot+05] use a stochastic dynamic optimization framework to compare centralized and decentralized investments decisions. For the centralized investment decisions, the maximization of social welfare is taken as objective, while for the decentralized case, a profit maximization is performed. Investments in two different technologies can be chosen. The electricity demand is modeled as a stochastic parameter using a recombining scenario tree. It is shown that the centralized as well as the decentralized problem formulation react similar to demand growth and demand uncertainty. However, the profit maximization tends to lead to lower investments than the social welfare maximization. This behavior is explained by the negative effect of new investments on electricity prices as well as by a price cap in the spot market which is below the VoLL.

The model described by Jensen and Meibom [JM08] is similar to the approach we are following. A fundamental electricity market model for the Nordic electricity market is used to determine the contribution margins for a stochastic dynamic investment model. Different uncertain events are included: the construction of an additional transmission line, the extension of the installed wind power capacity and high or low carbon prices. To keep the model computational tractable, only one investment alternative exists. The investor may invest in one combined heat and power gas turbine (combined cycle) at four predefined points in time. Two different cases are considered. In the first case, the investor does not hold any other plants in the market. In the second case, he already owns some power plants. It is shown that investors holding a portfolio of existing plants have a tendency to delay investments.

Hundt and Sun [HS09] use a linear mixed-integer, stochastic optimization model. A natural gas unit and a coal unit may be build in the period between 2008 and 2012. The fuel prices are modeled as a stochastic parameter. It is assumed that the logarithm of the natural gas price and the coal price are following a mean-reverting process. Correlated fuel price paths are created using a Monte Carlo simulation, then these price paths are reduced to a scenario tree with 16 different fuel price scenarios in 2012. It is assumed that the carbon prices are known for the whole period. Electricity prices are determined by a fundamental electricity market model considering short-term wind fluctuations using a recombining scenario tree. The main difference to our model can be found in the objective of the model. While our model focuses on a strategic, long-term horizon, the model presented in [HS09] focuses on a shorter time horizon.

The investment model we propose is similar to the models used in [HS09; JM08; Kle+08]. While all of these models consider fuel price or carbon price uncertainty, the consideration of uncertainty is limited to a few scenarios only. The objective of this thesis is to provide a framework that allows the consideration of a greater variety of stochastic fuel and carbon price scenarios. In the same time, electricity prices and electricity production of our model should be based on a fundamental electricity market model.

1.7 Summary

In this chapter, we gave an introduction to electricity markets. Thereby, we focused on the parts that are important for the valuation of power plants. We started with a description of different electricity markets. These markets are the electricity spot market, the derivatives market, the reserve market and the capacity market. For the investment alternatives we consider in this thesis, coal units and CCGT units, the most important markets are the markets for selling electricity, i.e., the spot and the derivatives market. Hence, we focus on these markets when we determine the contribution margin for new units. As forward electricity prices are assumed to be a predictor of future spot prices, we argue that it is sufficient to consider one of these markets. In the following parts of this thesis, we use the electricity spot market to derive electricity prices and production, as this is the market that is commonly modeled by fundamental electricity market models.

Furthermore, we discussed some peculiarities of electricity compared to other commodities. These peculiarities are the non-storability of electricity, the fact that supply must always match demand, the price-inelastic demand and its grid dependency. It is important to consider these peculiarities when electricity prices are modeled, as they are the reason for several characteristics of electricity spot prices: cycles, high-volatility, spikes, mean-reversion and negative prices.

In the next section, we described how electricity prices are determined. As most electricity market models are based on the assumption of perfect competition, we briefly introduced the microeconomic theory of pricing under perfect competition. It was shown that in a competitive market, prices correspond to the marginal costs of the most expensive unit required to satisfy the demand. Thereby, fixed costs are not considered. Next, we examined whether the assumption of perfect competition is pertinent for the German electricity market. Several studies indicate that the German electricity market was not fully competitive during the last years. Nevertheless, we argue that the assumption of perfect competition can be used to determine electricity prices, amongst others because the planned extension of transmission capacity between the European countries will increase competition.

In the following section, we reviewed the different uncertainties to which power plant investments are exposed. These uncertainties can be classified into volume risk, price risk and quantity risk. The main factors influencing these risks are fuel, carbon, and electricity price uncertainty as well as uncertainty about the legal framework. As the different risks are highly interrelated, it is important that this relationship is captured by the method used to determine future contribution margins for new units. We argue that a fundamental electricity market model is best suited for this purpose.

In the next section, we described different approaches to simulate fuel, carbon and electricity prices. There are a wide variety of different stochastic processes used for the simulation of these prices, and there is no consensus in literature which approach is appropriate. In our opinion, the difficulties related to fuel and carbon price modeling stress the importance to consider a wide variety of different scenarios.

In the following section, we presented two different methods used to evaluate power plant investments: the LCEP approach and the real options approach. The first one was mainly used in regulated market, and is not suited to consider the uncertainties to which generation companies are exposed in a deregulated market. The real options approach was developed for investments under uncertainty. Besides the question whether to invest or not, the real options approach also addresses the question when to invest. It is based on the assumption that the option to invest has a value, which must be considered when an investment opportunity is evaluated. The value of this option is related to new information that becomes available at later times. Due to its ability to consider uncertainties in an appropriate way, the investment models we describe in this thesis are based on the real options approach.

In the last section of this chapter, we gave an overview of related work. Thereby, we briefly described some simulation based and some game-theoretic approaches to power generation expansion, before we discussed in more detail the optimization based power generation expansion models. While there are some models that follow a similar approach as we do, combining the real options approach with a fundamental electricity market model, the consideration of fuel and carbon price uncertainty in these models is restricted to a few different scenarios. The focus of the investment models we propose in this thesis lies on the consideration of a wide variety of different fuel and carbon price scenarios.

Chapter 2

Fundamental Electricity Market Models

Fundamental electricity market models are—in our opinion—best suited to model electricity prices for the valuation of power plant investments. The reason therefore is that they can consider the interrelationship between fuel prices, carbon prices, the existing power generation fleet on the one side and electricity production and electricity prices on the other side.

In Section 1.4.5, we briefly outlined three different types of fundamental market models: the LDC model, the SDC model, and LP/MILP models. In this chapter, we describe these three models in more detail. Thereby, we focus on the LDC model and LP models, as these are the models we use in our investment models.

We start with a description of the LDC model. While its computational requirements are the lowest, it is also the most simplistic model. It does neither consider technical constraints, nor can it model uncertainty or price elastic demand. The inability of the model to consider technical constraints and uncertainty leads to an underestimation of prices in times of high demand, and an overestimation of prices in times of low demand. To account for these issues, we propose a function that increases electricity prices in times of high demand and decreases prices in times of low demand, while it does not change average prices. In addition, we use a mark-up function to consider disproportionately high prices in times of scarcity.

In the following section, we describe the SDC model. As the LDC model, it disregards technical constraints and uncertainty. The difference between the LDC and the SDC model is that the latter one can consider price elastic demand. In addition, the SDC model can use different merit-order curves, e.g., to consider different seasonal availabilities. If demand is assumed to be completely price in-elastic and a constant merit-order curve is used, then both models give the same results.

In the third section of this chapter, we focus on models based on an LP formula-tion of the electricity market. These models can consider technical constraints as well as short-term uncertainties like uncertainty of the feed-in from wind power or short-term demand uncertainty. A further advantage of LP based models is their ability to determine carbon prices endogenously. While these models are the most detailed, they also require the most computational effort. For power plant investment models, that require the electricity production and electricity prices for several years and different fuel price scenarios as input parameters, these models are computational intractable without simplifications.

For this reason, different simplifications are generally applied. Most of these sim-plifications are based on aggregation. For example, units of the same technology with similar constraints are often aggregated to unit groups. The aggregation of units with similar characteristics to unit groups makes the commonly used for-mulation of minimum up-time and minimum down-time constraints inapplica-ble. We propose an alternative formulation that also holds for unit groups, and is in addition more efficient than the formulation commonly used.

Another form of simplification is related to the optimization horizon. Instead of considering a whole year, only some typical days are often considered. However, also these simplifications have effects on the model, which are often disregarded in literature. The selection of some typical hours for the optimization has implica-tions on all intertemporal constraints like the minimum up-time constraints. Espe-cially the constraints related to the storage operation have to be adjusted to guar-antee an—at least approximately—feasible storage operation. We discuss these issues and propose two alternative formulations that take these problems into ac-count.

In the last part of this chapter, we compare the prices obtained with the LDC and the LP model with prices observed at the EEX. We select these two types of models for the comparison, as these are the models we use in our investment models in the following chapters. We choose the LDC model, as it is the fastest model. The LP-based model is much slower, but has the advantage that it offers a possibility to determine carbon prices endogenously. The comparison shows that average EEX prices are relatively well approximated, especially by the LDC model with the adjustments we propose. However, the standard deviation of model prices is significantly below the standard deviation of EEX prices. Nevertheless, we argue that these models can be used for power plant valuation, as a good approximation of average prices is more important for the valuation of mid-load or base-load units than the approximation of the standard deviation.

2.1 Load–Duration Curve Models

The simplest electricity market model, which can be used to determine electricity prices and the electricity production, is based on the LDC. While its computational requirements are low, it has the drawback that it cannot consider technical restrictions and uncertainty. As a consequence, model prices exhibit a lower variance compared to observed electricity prices. Especially in times of scarcity, model prices are significantly below observed prices. For that reason, some models use scarcity mark-ups to adjust model prices.

In the following, we first introduce the basic LDC model. Next, we present two functions that can be used to adjust model prices. The first one is a mark-up function that is applied in times of scarcity, while the second function is used to increase the variance of the prices. Thereby, it aims to simulate the effect of start-up costs and avoided start-up costs, respectively.

2.1.1 Basic Model

The LDC model determines electricity production and electricity prices based on the LDC and the merit-order curve. On the top right of Figure 2.1, a typical LDC is shown. The LDC corresponds to the electrical load for a given period sorted in descending order of magnitude. In electricity markets where electricity production from renewable sources must be fed-in, like it is the case in Germany, this production is in general directly deduced from the load.

The merit-order curve shows the available capacity sorted in ascending order of marginal costs. Based on the assumption of perfect competition, under which operators bid corresponding to their marginal costs, the merit-order curve corresponds to the supply curve.

Under the assumption that the cheapest units are used to satisfy the demand, the price–duration curve can be derived, as it is shown in Figure 2.1. Starting from the LDC (the curve on the top right), the corresponding electricity prices for a given duration are determined by going to the (rotated) supply curve on the top

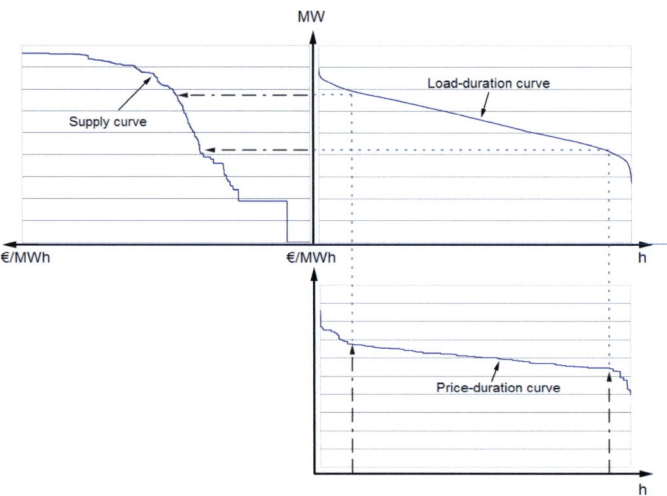

Figure 2.1: Determination of the price–duration curve based on the LDC and the supply curve.

left. The price of the supply curve and the duration of the LDC define a point of the price–duration curve. The complete price–duration curve can be derived by repeating this procedure for every point of the LDC.

The main advantage of LDC models is the fast computation. However, there are several disadvantages:

- No consideration of uncertainty like stochastic feed-in from renewables or short-term demand uncertainty.

- No consideration of technical restrictions. For example, it is assumed that units operate with a constant efficiency. Operational constraints like minimum up-times or minimum down-times are neglected.

- No consideration of start-up costs.

- It is assumed that the merit-order curve is the same for the whole period. This implies a constant availability of the power plants as well as constant fuel prices within the considered period.

- No consideration of price-elastic demand.

- Pumped-storage operation cannot be considered realistically.[1]

_____

[1]A possibility to approximate storage operation is to represent pumped-storages through increased demand in periods of low demand. During these periods, prices are low, and storages may

Despite these disadvantages, the application of the LDC to determine electricity prices is relatively popular, especially in cases where a fast computation is important. Examples for the usage of the LDC in power generation expansion models are Klein et al. [Kle+08] and Olsina [Ols05].

2.1.2 Adjustments to the Basic Model

The before mentioned limitations of the LDC model have several effects on model prices, which we discuss in the following. Then, we propose two different functions to adjust model prices to get more realistic results.

The effects of the limitations of the LDC model are:

- The non-consideration of start-up costs leads to an underestimation of electricity prices. This effect is particularly distinct during short periods of high load, when peak load units have to be started to satisfy electricity demand. As peak load units operate only during a short period, high prices occur because the whole start-up costs must be gained in this short period.

- The disregard of minimum down-times results in an overestimation of available capacity. Hence, model prices underestimate real prices.

- The non-consideration of minimum up-times has the opposite effect. If a unit cannot be shut-down, electricity prices may be below the marginal costs of such a unit.

- A similar effect can be observed when start-up costs can be avoided by not shutting-down a unit. In such a case it might be cheaper to produce electricity even if the marginal costs of the unit are higher than the electricity price. One may think that such situations can only occur during low load periods, e.g., during the night. However, such a situation can also occur during high-load periods in case of unforeseen high feed-in from intermittent sources like wind power.

- The assumption of a constant merit-order curve has several effects. First, the variance of model prices is lower than in reality, because model prices are

---

be charged. The production of electricity by pumped-storages can be included by adding storage capacity to the merit-order curve at high price levels. However, the determination of accurate price-levels is a non-trivial task (for different possibilities, see Burger et al. [Bur+07, p. 171-173]). Furthermore, this representation of storage operation may violate storage content restrictions, e.g., storages may produce more electricity than they have stored.

based on average fuel and carbon prices. The real price curve can significantly fluctuate within the considered period. For example, the EEX spot price for natural gas increased from 3.50 €/GJ on July 02, 2007 to 6.61 €/GJ on the December 28, 2007.[2]

Second, during seasons of high demand, model prices tend to be higher than observed prices, while for seasons with low demand, it is the other way around. The reason for this behavior is that maintenance of plants is generally scheduled during seasons of low demand. Hence, the availability is below average during this periods, and above average during periods of high demand, when no maintenance is scheduled.

- Owing to short-term uncertainties, an increased number of units must operate in part-load to be able to react on short-term fluctuations. If the feed-in from wind power is significantly below the predicted values, new units must be started to balance these deviations. Hence, uncertainty increases electricity prices. The effects of uncertainty can only be partly considered in LDC models. One possibility is to assume lower efficiencies for units due to the less efficient part load operation. Another possibility is to reduce the availability of units by the amount of required reserve power, or to add the reserve power to electricity demand.

Deviations between model prices and observed prices are often reported in literature (e.g., Hirschhausen et al. [Hh+07], Möst and Genoese [MG09]). In general, these deviations are the greatest in times of scarcity. The explanations of these deviations are different and range from abuse of market power [Hh+07] over investment cost recovery [MG09] to data imprecisions (Swider et al. [Swi+07]).

To account for such deviations, Möst and Genoese use a piecewise constant mark-up function to adjust electricity prices. Based on the scarcity factor, which is the available supply divided by the demand, they define five different intervals for a scarcity factor below 2.0. If the scarcity factor is in one of these five intervals, a constant share of a unit's fixed costs is added to the unit's marginal costs. The lower the scarcity factor is, the higher is the share of the unit's fixed costs that is added.

In the following, we propose similar adjustments, which are motivated by the results shown in Figure 2.2. This figure shows deviations between observed EEX

---

[2]Note that even if generation companies purchase their fuel on long-term contracts with fixed costs, fluctuating spot market prices might create opportunity costs which have to be included in the electricity price.

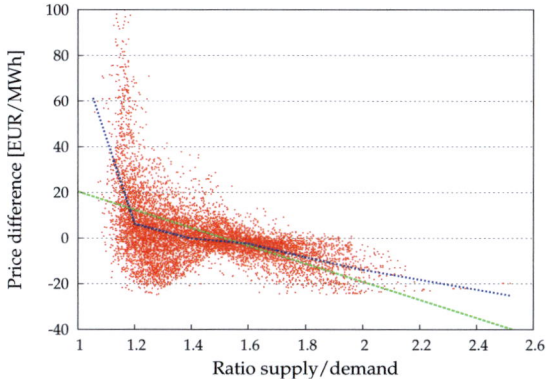

Figure 2.2: Deviations between EEX prices and model prices as a function of the available capacity divided by the demand and possible linear and piecewise-linear regression lines.

prices and model prices depending on the available capacity divided by the demand, i.e., the scarcity factor. Besides these deviations, two trendlines are included, a linear one and a piecewise linear one. A perfect model would produce points which all lie on a horizontal line with price deviations of zero. Most of the reasons why this optimal result is not achieved have been discussed at the beginning of this subsection, e.g., the non-consideration of technical constraints.

Based on the reasons for price deviations we mentioned so far, one would expect the model to underestimate prices in case of scarcity, while real prices might be lower in times of low demand. The deviations should be around the trendlines plotted in Figure 2.2. However, negative price deviations can also be observed when supply is relatively scarce. Some of these deviations can be explained by the non-consideration of pumped-storages and hydro-reservoirs. Other reasons for these deviations are the non-consideration of uncertainty, an underestimation of the available capacity during these periods, and possibly too high fuel prices. But it should also be mentioned, that some of the deviations may be due to data inaccuracy. For example, the exact efficiencies are not known for all plants, neither are the hourly availabilities of the plants.

A straightforward approach to adjust electricity prices would be to use either the linear or the piecewise-linear trendline depicted in Figure 2.2. Depending on the scarcity factor, the corresponding price difference would be added to the model's prices. These functions—especially the piecewise linear function—provide good approximations of electricity prices for the period we examined, the years 2006–

2009. However, there is a problem with these functions, which makes them unsuitable, at least in the form they are shown in Figure 2.2. Under certain circumstances, the application of these adjustments can result in average electricity prices that are below the marginal costs of a unit. Such a behavior may be realistic in the short-term due to technical restrictions, but no operator sells electricity during longer time-periods below marginal costs.

This problem is caused by the introduction of negative price adjustments. In our tests, these negative price adjustments significantly improve electricity price forecasts in periods of low prices. While the linear and piecewise-linear functions depicted in Figure 2.2 work well for electricity markets with a similar ratio of supply and demand as those in the years between 2006 and 2009, an increase of the supply/demand ratio leads to the mentioned problem. In such a case, the number of negative price adjustments increases and average prices decrease. The consequence can be negative contribution margins for some units.[3]

To circumvent this problem, we use two different types of adjustments: a linear function that accounts for effects that are mainly due to the non-consideration of technical constraints, and a piecewise-linear function that represents mark-ups in case of scarcity.

The first function we use balances two opposing effects: lower model prices during peak load periods, and higher model prices during periods of low load. As discussed before, the reason for these deviations can be the non-consideration of technical constraints like start-up costs or minimum down-times respectively the non-consideration of avoided start-up costs or minimum up-times. To avoid that units operate (on average) below marginal costs, we define this function such that it does not change average electricity prices. For this purpose, we use the following function to calculate price adjustments (in €/MWh):

$$\text{price adjustments}_t = \left(\frac{\text{load}_t}{\text{average annual load}} - 1 \right) \cdot 50. \qquad (2.1)$$

In periods of low load, the first factor is negative and prices are decreased. If load is above average, the first factor is positive and prices are increased. The value of 50 was obtained by linear regression based on the comparison of model prices and EEX prices for 2006 and 2007. Note that this formulation guarantees that the contribution margin of a unit does not get negative. The reason is that

---

[3]These contribution margins consider only variable costs. Therefore, negative contribution margins are not realistic.

the cheapest units are always committed first. If a unit produces during periods of low load (and prices are decreased), it also produces during all hours when load is above average (and prices are increased). As function (2.1) is defined such that the sum of the positive adjustments corresponds to the sum of the negative adjustments, a unit that produces during all periods of price increases cannot get a lower contribution margin than without these adjustments.

The second function we use to adjust model prices is related to the scarcity of supply. These mark-ups can be interpreted as exertion of market power or fixed cost recovery. Consequently, the mark-ups calculated by this functions are greater or equal to zero. We use the following formula to determine these mark-ups (in €/MWh):

$$\text{price adjustments}_t = \begin{cases} 91 - 70 \cdot \frac{\text{supply}_t}{\text{load}_t} & \text{if } \frac{\text{supply}_t}{\text{load}_t} < 1.3, \\ 0 & \text{otherwise.} \end{cases} \tag{2.2}$$

The values chosen for the adjustments have been found by linear interpolation between deviations of EEX prices and those of the LDC model adjusted following function (2.1) for the years 2006 and 2007.

Note that we use constant plant availabilities. Therefore, the supply in equation (2.2) is constant. Combining functions (2.1) and (2.2) results in a piecewise-linear adjustment function depending on the (residual) load.

2.2 Supply–Demand Curves Model

The determination of electricity prices based on SDCs is similar to the LDC model. Owing to the faster running times of the LDC model, we opt to use an LDC model to determine the contribution margins for the new units in our investment models. Therefore, we only briefly outline the SDC model in the following.

SDC models determine electricity prices as the intersection of the merit-order curve and the demand curves for different moments in time. This is illustrated in Figure 2.3. The figure shows three different demand curves (the vertical lines), each representing the demand at a different moment in time. The left demand curve is price inelastic, while the other two are examples of price-elastic demand curves. All units that are left of the intersection between the supply and the demand curve produce electricity, all units right of the intersection are not used. To reduce the number of calculations, a demand curve may represent several hours.

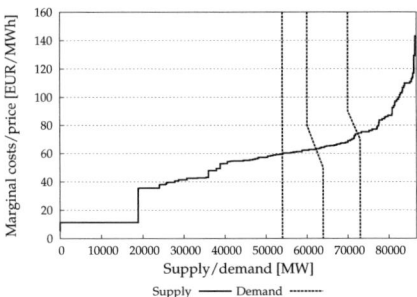

Figure 2.3: Price determination based on the SDC. One supply curve and three different demand curves are shown.

Compared to the LDC model, the SDC model has two advantages. First, price-elastic demand can be considered. Second, different supply curves can be used to represent different seasonal plant availabilities or fuel prices. However, as the LDC model, the SDC model is not able to consider technical constraints, nor can it represent storage operation in an appropriate way. As multiple demand curves (and possibly supply curves) are used, the SDC model is slower than the LDC model.

If no price-elastic demand and only one supply curve are used, the results of SDC models correspond to the results of LDC models. An example for the usage of an SDC model in the context of power generation expansion is Botterud et al. [Bot+05].

2.3   Linear Programming Models

The main disadvantage of LDC models as well as SDC models is the non-consideration of technical constraints of power plants. To be able to consider technical constraints, the power generation can be described as an MILP problem. In such a model, continuous variables are used to represent the electricity production of the different units. To be able to consider technical constraints like reduced part-load efficiency, binary variables are required to model the status (on/off) of the units.[4] Depending on the size of the considered market, MILP models are computational intractable. In general, LPs can be solved faster than MILPs. Therefore, it is a common technique to solve the continuous relaxation of the problem instead of

---

[4]In the (common) case of unit aggregation, integer variables are used instead of binary variables.

the MILP (see Weber [Web05]). However, it should be noted that by replacing the MILP with an LP, some constraints—like reduced part-load efficiency—can only be approximated.

Modeling an electricity market as MILP or LP is quite popular, see, e.g., Leuthold et al. [Leu+08], Meibom et al. [Mei+06], Müsgens [Müs06], Ravn et al. [Rav+01], Weigt and Hirschhausen [WH08]. As a comparison done by the author of this thesis [Bar+08] between an MILP description of the German electricity market and its LP relaxation shows similar results for both models, we use an LP model.

In this section, we present several LP formulations of fundamental electricity market models with different levels of detail. We start with a simple deterministic model, before we introduce some stochastic elements based on a scenario tree. Depending on the level of detail modeled, it is computationally intractable to solve such models for a longer time period (like a year for example), even with efficient commercial solvers.[5] To circumvent such problems, a common approach is to represent a longer time period by some typical days (e.g., Hundt and Sun [HS09], Sun et al. [Sun+08], Swider [Swi07], Swider and Weber [SW07]). Compared to normal models considering a complete period, these models require some extra constraints, which are often overlooked in literature. After discussing these additional constraints, we present a method based on a recombining scenario tree that allows to add stochasticity to the model without adding too much computational burden. Finally, we show how electricity prices and carbon prices can be determined using an LP-based electricity market model.

2.3.1 Basic Deterministic Model

The basic deterministic model described in this subsection is the basis for the other models presented afterwards. In all of our models, we only consider the electricity production. The heat production is disregarded. Furthermore, we assume that there is enough transmission capacity, so that we do not have to model the transmission network. As a consequence, we do not have to consider different regions. Electricity exchange with regions outside of the model is assumed to be known and subtracted from or added to the demand. Finally, we assume that electricity demand is completely price-inelastic. All variables mentioned in our models are non-negative continuous variables.

---

[5]To solve the LP models described in this thesis, we use ILOG CPLEX 11.2.

Objective Function:

The objective function of a fundamental electricity model is the minimization of the total costs TC over the whole optimization horizon. The total costs for each unit are composed of three different types of costs: the costs for the fuel consumption FC including carbon costs, operating and maintenance costs OC and start-up costs SC. Let $\mathcal{U}$ be the set of all units in the market, and $\mathcal{T}$ the set of all timesteps. Then, the objective function can be defined as:

$$\min TC = \sum_{u \in \mathcal{U}} \sum_{t \in \mathcal{T}} (FC_{ut} + OC_{ut} + SC_{ut}). \tag{2.3}$$

Fuel Costs:

Thereby, the fuel consumption costs are defined as the product of the fuel consumption F and the associated costs. These costs consist of the fuel price $c^{\text{Fuel}}_{f(u)}$ (with $f(u)$ denoting the fuel used by the unit u) and carbon costs. The carbon costs are obtained by multiplying the CO_2 costs per tonne c^{CO_2} by a factor $\kappa_{f(u)}$ describing the amount of emissions produced depending on the fuel. We assume that the fuel prices and the carbon price remain constant within the considered period.

$$FC_{ut} = F_{ut} \cdot (c^{\text{Fuel}}_{f(u)} + c^{CO_2} \cdot \kappa_{f(u)}) \qquad \forall u, t. \tag{2.4}$$

We make the simplifying assumption that the fuel consumption depends linearly on the produced electricity P. To model lower part-load efficiency, a fixed amount of fuel consumption is added if a unit is online. As we use an LP model, we follow the approach proposed by Weber [Web05] and use the continuous variable P^{Online} to describe the capacity (in MW) currently online.[6] The parameters α and β can be determined depending on the full-load efficiency, the part-load efficiency, and the maximum and minimum production of a plant.

$$F_{ut} = \alpha_u \cdot P^{\text{Online}}_{ut} + \beta_u \cdot P_{ut} \qquad \forall u, t.$$

---

[6]Using an MILP, one would use instead of P^{Online} a binary variable U^{Online}, which indicates whether a plant is online or not. To transform our continuous model into an MILP model, one has to replace P^{Online} with $U^{\text{Online}} \cdot p^{\text{max}}$, whereby p^{max} denotes the maximum capacity of the plant. One should note that, compared to the MILP, the LP formulation leads to an underestimation of the costs. The reason is that one may decide to start only a fraction of a unit. Thereby, two effects occur. First, the start-up costs are lower, as these costs depend on the amount of capacity started. Second, the unit is assumed to operate under full-load efficiency instead of part-load efficiency.

Operating and Maintenance Costs:

In addition to the fuel consumption costs, there are costs for the operation and maintenance of a plant. We assume that these costs depend solely on the electricity production. The factor γ denotes the operating and maintenance costs in €/MWh.

$$OC_{ut} = \gamma_u \cdot P_{ut} \qquad \forall u, t.$$

Start-Up Costs:

The start-up costs are composed of two different parts: additional fuel usage and abrasion. The parameter θ denotes the additional fuel usage in $\text{MWh}_{\text{fuel}}/\text{MW}_{\text{elec}}$, c^{abrasion} are the costs in €/MW calculated for abrasion. The capacity that is started-up is denoted P^{StartUp}.

$$SC_{ut} = (\theta_u \cdot (c_{f(u)}^{\text{Fuel}} + c^{CO_2} \cdot \kappa_{f(u)})) \cdot P_{ut}^{\text{StartUp}} \qquad \forall u, t.$$

Power Balance Constraint:

The power balance constraint ensures that supply matches demand at every moment in time. The sum of electricity produced by all units must correspond to the electricity load plus the electricity W used to charge pumped-storages. As we assume that electricity production by renewable resources (wind power, run-of-river plants) must be fed-in, we deduce this production directly from the load. The difference between the load and the production by renewable sources is denoted with l.

$$\sum_{u \in \mathcal{U}} P_{ut} = l_t + \sum_{u \in \mathcal{U}^{\text{Storage}}} W_{ut} \qquad \forall t. \tag{2.5}$$

Reserve Constraints:

As explained in Section 1.1.1, three different types of reserve energy exist in Germany. However, from a modeling point of view, primary and secondary control power are similar. As it is done by many fundamental market models (e.g., Burger et al. [Bur+07], Meibom et al. [Mei+06]), we consider just one type of negative reserves, $P^{\text{Res}-}$. Concerning positive reserves, we distinguish spinning reserves $P^{\text{ResSpin}+}$ and non-spinning reserves $P^{\text{ResNonSpin}+}$. Spinning reserves must be provided by units that are already online. Non-spinning reserves may also be provided by units that are offline, if they can start-up fast enough. We assume that gas-turbine units fulfill this requirement. Other units may also contribute to the non-spinning reserves in our model, but they must be online to do so. For each type of reserves, a given demand ($d_t^{\text{Res}-}$, $d^{\text{ResSpin}+}$, and $d^{\text{ResNonSpin}+}$, respectively) must be satisfied.

$$\sum_{u\in\mathcal{U}} P_{ut}^{\text{Res}-} = d_t^{\text{Res}-} \qquad \forall t,$$

$$\sum_{u\in\mathcal{U}} P_{ut}^{\text{ResSpin}+} = d_t^{\text{ResSpin}+} \qquad \forall t,$$

$$\sum_{u\in\mathcal{U}} P_{ut}^{\text{ResNonSpin}+} = d_t^{\text{ResNonSpin}+} \qquad \forall t.$$

Production Constraints:

The maximum production of a unit is restricted by the capacity that is online. Furthermore, if a unit provides positive reserve energy, this amount must be subtracted from the maximum production. For units that do not start fast enough to provide positive non-spinning reserves when they are offline, the following constraint is used:

$$P_{ut} + P_{ut}^{\text{ResSpin}+} + P_{ut}^{\text{ResNonSpin}+} \leq P_{ut}^{\text{Online}} \qquad \forall u \notin \mathcal{U}^{\text{FastStart}}, t. \qquad (2.6)$$

For units that are able to provide non-spinning reserves when they are offline, we use the following constraint:

$$P_{ut} + P_{ut}^{\text{ResSpin}+} \leq P_{ut}^{\text{Online}} \qquad \forall u \in \mathcal{U}^{\text{FastStart}}, t.$$

Most plants have a minimum production level. The following inequality ensures that electricity production does not drop below this minimum production level. Thereby, $p^{\min}$ denotes the minimum output relative to the capacity online. Similar to the maximum production constraint, the supply of negative reserves must be considered.

$$P_{ut} + P_{ut}^{\text{Res}-} \geq p_u^{\min} \cdot P_{ut}^{\text{Online}} \qquad \forall u, t.$$

The capacity online is restricted by the maximum capacity $p^{\max}$ of a unit. As we do not explicitly model plant outages, we reduce the maximum capacity of a plant by multiplying it with typical availability rate ρ. If a longer time horizon is considered, ρ may depend on the timestep.

$$P_{ut}^{\text{Online}} \leq p_u^{\max} \cdot \rho_{ut} \qquad \forall u, t.$$

For fast-starting units, we have to ensure additionally that the sum of electricity produced and positive reserves provided does not exceed maximum available capacity (for all other units, the maximum production constraint (2.6) guarantees this requirement):

$$P_{ut} + P_{ut}^{\text{ResSpin}+} + P_{ut}^{\text{ResNonSpin}+} \leq p_u^{\max} \cdot \rho_{ut} \qquad \forall u \in \mathcal{U}^{\text{FastStart}}, t.$$

The capacity started-up must be greater or equal to the difference of the capacity online at the moment and the capacity online during the previous timestep.[7]

$$P_{ut}^{\text{StartUp}} \geq P_{ut}^{\text{Online}} - P_{u,t-1}^{\text{Online}} \qquad \forall u, t. \qquad (2.7)$$

We also define the capacity shut-down P^{ShutDown}, which we need later for the minimum down-time constraints. The reasoning in the definition is the same as for P^{StartUp}.[8]

$$P_{ut}^{\text{ShutDown}} \geq P_{u,t-1}^{\text{Online}} - P_{ut}^{\text{Online}} \qquad \forall u, t. \qquad (2.8)$$

Minimum Up- and Down-Time Constraints:

Once a unit is started, it has to stay online for a certain time. A common formulation for minimum up-time constraints for an MILP formulation of the unit commitment problem is the following (e.g., Carrión and Arroyo [CA06]):

$$\sum_{\tau=t}^{t+t_u^{\text{MinUp}}-1} U_{u\tau}^{\text{Online}} \geq t_u^{\text{MinUp}} \cdot (U_{ut}^{\text{Online}} - U_{u,t-1}^{\text{Online}}) \qquad \forall u, t, \qquad (2.9)$$

with U^{Online} being a binary variable equal to one if the unit is online and zero otherwise. If a unit is started at timestep t, the right hand side of the inequality gets positive. In such a case, the unit must stay online for at least its minimum up-time. This is ensured by summing up the status variables for the next $t^{\text{MinUp}} - 1$ timesteps[9] on the left hand side of inequality (2.9).

---

[7]Formally defined, the following equation must hold for the capacity started: $P_{ut}^{\text{StartUp}} = \max(0; P_{ut}^{\text{Online}} - P_{u,t-1}^{\text{Online}})$. Inequality (2.7) defines only a lower bound for P^{StartUp}. However, as we consider a cost minimization problem, and P^{StartUp} is (indirectly over the term SC) included in the objective function, this formulation is sufficient.

[8]However, we have no costs associated with shutting-down units. Hence, inequality (2.8) is not sufficient to guarantee that $P_{ut}^{\text{ShutDown}} = \max(0; P_{u,t-1}^{\text{Online}} - P_{ut}^{\text{Online}})$ holds. Therefore, we add relative low costs for shutting down units to the objective function. For a better readability, these costs are not shown in the objective function (2.3).

[9]As the current timestep t is included, the -1 is required.

time	10	11	12	13	14
units online	2	2	4	2	1

Table 2.1: Example for an infeasible schedule not detected by constraint (2.10).

However, there are circumstances under which restriction (2.9) does not work. This is the case if either similar units are aggregated to unit groups, or if the binary variables U^{Online} are replaced by continuous variables as it is done for LP formulations.

This problem is often disregarded in literature. For example, Weber [Web05] proposes a constraint similar to (2.9) for his LP model:

$$P_{ut}^{Online} - P_{u,t+1}^{Online} \leq P_{u\tau}^{Online} \qquad \forall u, t, \forall \tau \in (t - t_u^{MinUp}; t]. \qquad (2.10)$$

This constraint is also used for the LP model by Meibom et al. [Mei+06]. Let us briefly demonstrate with a simple example why this formulation does not work in LP models or models using unit aggregation, before we propose an alternative formulation, which is also more efficient.

Example. *Assume there are four units aggregated to one unit group. The minimum up-time of these units is three hours. Table 2.1 shows a schedule which is infeasible. At noon, two additional units are started. These units must stay online until at least 14 o'clock. Applying inequality (2.10), we get:[10]*

$$\left. \begin{array}{l} P_{13}^{Online} - P_{14}^{Online} \leq P_{13}^{Online}, \\ P_{13}^{Online} - P_{14}^{Online} \leq P_{12}^{Online}, \\ P_{13}^{Online} - P_{14}^{Online} \leq P_{11}^{Online}. \end{array} \right\} \Leftrightarrow \left\{ \begin{array}{l} 2 - 1 \leq 2, \\ 2 - 1 \leq 4, \\ 2 - 1 \leq 2. \end{array} \right.$$

It can be seen that the constraint that should guarantee the minimum up-time restriction fails. The reason for this is that the constraint is unable to detect that some units are already shut-down at 13 o'clock. If no units are shut-down at 13 o'clock, constraint (2.10) detects the infeasible schedule.

---

[10]Following the way the restriction is defined in [Web05], we must additionally check for $P_{13}^{Online} - P_{14}^{Online} \leq P_{10}^{Online}$. That is probably a typo, as otherwise one would require a minimum up time of $t^{MinUp} + 1$.

To avoid this problem, we propose to ensure that the capacity online is always greater as or equal to the sum of the capacity started during the previous $t^{\mathrm{MinUp}} - 1$ periods. Formally defined, this is:

$$P_{ut}^{\mathrm{Online}} \geq \sum_{\tau = t - (t_u^{\mathrm{MinUp}} - 1)}^{t} P_{u,\tau}^{\mathrm{StartUp}} \qquad \forall u, t. \qquad (2.11)$$

Example (continued). *Applying the proposed inequality (2.11) to the simple example presented before, the infeasibility of the schedule in Table 2.1 is detected by the following restriction:*

$$P_{14}^{\mathrm{Online}} \geq P_{12}^{\mathrm{StartUp}} + P_{13}^{\mathrm{StartUp}} + P_{14}^{\mathrm{StartUp}} \Leftrightarrow 1 \geq 2 + 0 + 0 \quad \notz.$$

We define the minimum down-time constraints in a similar manner. The capacity that is offline at the moment must be greater as or equal to the sum of the capacity shut-down during the $t^{\mathrm{MinDown}} - 1$ previous periods:

$$P_u^{\mathrm{max}} - P_{ut}^{\mathrm{Online}} \geq \sum_{\tau = t - (t_u^{\mathrm{MinDown}} - 1)}^{t} P_{u,\tau}^{\mathrm{ShutDown}} \qquad \forall u, t. \qquad (2.12)$$

Empirical tests have shown that using restrictions (2.11) and (2.12) instead of (2.10) and the corresponding minimum down-time constraint is much more efficient. The average runtime of a fundamental electricity market model of Germany with the new constraints is between 16% and 55% lower than those with the old constraints (see Geiger et al. [Gei+08]).

Storage Constraints:

The content of the pumped hydro storages V (in MWh) at the current timestep is equal to the content at the previous timestep minus the production at the current timestep plus the electricity used for charging the storage multiplied by the charging efficiency η.

$$V_{ut} = V_{u,t-1} - P_{ut} + \eta_u \cdot W_{ut} \qquad \forall u \in \mathcal{U}^{\mathrm{Storage}}, t. \qquad (2.13)$$

To avoid that the model considers the initial storage content as free energy and completely empties storages at the end of the planning horizon, there are two possibilities. The first one is to associate some value to the final storage content in the objective function (see, e.g., Meibom et al. [Mei+06]). The second possibility is to require that the final storage content is equal to the initial content (e.g., Swider [Swi07]). As the determination of appropriate values for the storage content is a non-trivial task, we follow the latter approach.

$$V_{u,T} = V_{u,0} \qquad \forall u \in \mathcal{U}^{\text{Storage}}.$$

Similar to the electricity production, there is an upper bound for the storage charging, w^{max}, and the storage content, v^{max}:

$$V_{ut} \leq v_u^{\text{max}} \qquad \forall u \in \mathcal{U}^{\text{Storage}}, t,$$
$$W_{ut} \leq w_u^{\text{max}} \qquad \forall u \in \mathcal{U}^{\text{Storage}}, t.$$

CO_2 Constraints:

The method commonly used to consider CO_2 restrictions in electricity market models is to include the CO_2 costs in the objective function, as it is done in constraint (2.4) defining the fuel costs. To be able to apply this method, carbon price forecasts are required. As discussed in Section 1.4.3, carbon price forecasting is quite difficult.

Instead of adding a penalty for emitted CO_2, another possibility is to restrict the maximal amount of allowed emissions in the model. For this purpose, the following constraint can be used:

$$\sum_t \sum_u F_{ut} \cdot \kappa_{f(u)} \leq em^{\text{max}}, \tag{2.14}$$

where em^{max} corresponds to the maximum allowed tons of CO_2 in the considered period. This value can be taken from the National Allocation Plans (NAP), in which each state participating in the EU ETS decides on the allowed emissions for the different industry sectors.

Using constraint (2.14) to limit the amount of emissions is too strict compared to reality. There are different possibilities to get additional emission allowances. They can either be bought from abroad or other industry sectors. Alternatively, some non-used allowances from another year (of the same trading period) may be used. Nevertheless, equation (2.14) is valuable, as it allows to determine the value of emission allowances for the considered period. This approach is presented in Section 2.3.5.

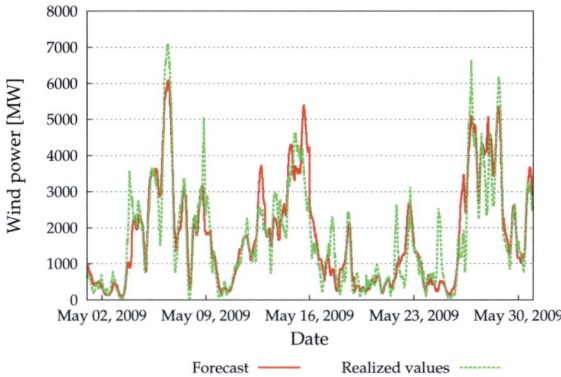

Figure 2.4: Wind power forecasts and realized wind power [MW] in the transpower control area in May 2009. Source: transpower.

2.3.2 Stochastic Model

The model described in the previous subsection is based on the assumption that all input parameters are known with certainty. However, this assumption is not realistic. An example for a stochastic parameter is the production from wind power. Figure 2.4 shows the predicted and the observed wind power production in the transpower control area, one of the four control areas in Germany. While the general trend is well approximated by the forecasts, there are sometimes deviations of over 2,000 MW.

To account for these uncertainties, electricity market models can be formulated as stochastic programming problems. In the following we briefly outline the main ideas of stochastic programming. For a general introduction to stochastic programming, see, e.g., Birge and Louveaux [BL97] or Kall and Wallace [KW94].

Stochastic programming considers uncertainty usually in the form of a scenario tree. An example for a three-stage scenario tree is shown in Figure 2.5. A stochastic programming problem consists of at least two stages. Each stage can comprise one or several timesteps. The first stage is a deterministic stage, in which all parameters are known with certainty. The succeeding stages are stochastic, which means that there are different possible outcomes for the uncertain parameters.

Let $\mathcal{N}$ be the set of nodes of the scenarios tree. Each node in Figure 2.5 corresponds to one realization of the stochastic parameter. If wind power is modeled as stochastic parameter, each node represents a different wind power scenario. It should be noted that even if two nodes in the scenario tree (e.g., node 4 and 6)

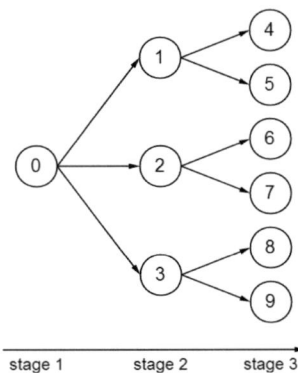

Figure 2.5: Illustration of a scenario tree used for stochastic programming.

represent the same outcome of a scenario (in our case wind power production), they cannot be combined if the problem contains intertemporal constraints. An example for such a constraint is the storage content constraint (2.13).[11]

The arcs between the nodes represent possible transitions from one scenario to another. Based on the transition probabilities $\psi_{n' \to n}$ between the nodes, the occurrence probability π_n can be determined for each node n.

To account for the different scenarios, an additional index must be added to all variables that are influenced by the stochastic parameter (or that can react on the variations of the parameter). Depending on the size of the scenario tree, this increases the problem size significantly.

Extending our deterministic electricity market model to a stochastic one with the wind power production as stochastic parameter is straightforward. Every variable (with a possible exception for the variables used to provide reserve energy[12]) depends additionally on the scenario, and every constraint must hold for all scenarios. In the following, only the constraints that substantially differ from the deterministic version are described.

---

[11]The reason why nodes cannot be combined is the following: Assume that the wind power production for node 4 and 6 is the same, while the one for the previous nodes 1 and 2 is different. Therefore, electricity production of a pumped-storage in node 1 may be different from the one in node 2. As a consequence, also the storage content is different. As the current storage content depends on the previous content, the storage content for node 4 and 6 may be different, even if the electricity production of the storage is the same at this stage. The problem is a path-dependent problem.

[12]In Germany, bids for minute reserves must be made until 10 a.m. for the succeeding day. Therefore, the supply of minute reserves is fixed at least 14 hours in advance. The supply of primary and secondary control power is even fixed for a whole month. If the supply is fixed in advance, it must be the same for all scenarios.

The different parts of the objective function remain the same, but the costs for all possible scenarios must be multiplied by their occurrence probability:

$$\min TC = \sum_{u \in \mathcal{U}} \sum_{t \in \mathcal{T}} \sum_{n \in \mathcal{N}} \pi_n \cdot (FC_{utn} + OC_{utn} + SC_{utn}).$$

As we subtract wind power directly from electricity demand, the residual electricity demand becomes scenario dependent. The power balance constraint is now:

$$\sum_{u \in \mathcal{U}} P_{utn} - \sum_{u \in \mathcal{U}^{\text{Storage}}} W_{utn} = l_{tn} \qquad \forall t, n.$$

The minimum up-time restriction can be written as

$$P_{utn}^{\text{Online}} \geq \sum_{\tau = t - (t_u^{\text{MinUp}} - 1)}^{t} P_{u\tau n'}^{\text{StartUp}} \qquad \forall u, t, n.$$

Thereby, n' is the corresponding predecessor node of n at timestep τ. The minimum down-time restriction is

$$P_u^{\max} - P_{utn}^{\text{Online}} \geq \sum_{\tau = t - (t_u^{\text{MinDown}} - 1)}^{t} P_{u\tau n'}^{\text{ShutDown}} \qquad \forall u, t, n.$$

The storage content is now restricted by the following constraint

$$V_{utn} = V_{u,t-1,n'} - P_{utn} + \eta_u \cdot W_{utn} \qquad \forall u \in \mathcal{U}^{\text{Storage}}, t, n.$$

2.3.3 Model Based on Typical Days

Both models presented in the previous subsections are based on the assumption of consecutive timesteps. For our investment module, we are interested in power market results for a whole year. Modeling a whole year is computationally intractable. A common approach to circumvent such problems is to consider just a few characteristic days (e.g., Sun et al. [Sun+08], Swider [Swi07], Swider and Weber [SW07]). In [Sun+08; Swi07; SW07], every two months of a year are represented by one workday and one weekend day. Each of these days consists of twelve typical load profiles with a duration of two hours each.

While the representation of the year by typical days significantly reduces the computational burden, it requires some model modifications. We discuss these modifications in the following for the basic deterministic model.

Objective Function:

As there are more workdays than days of the weekend, the different timesteps in the model do not occur equally often. Hence, they must be accordingly weighted. This can be done by introducing the weighting factor f_t.[13] If several hours are aggregated to one timestep, then an additional parameter d_t indicating the duration of each timestep is added. It is important to notice that only the fuel costs and the operating costs have to be multiplied by d_t. The start-up costs are independent of the duration of a timestep, a fact that is often not considered (e.g., [Swi07; SW07][14]). The objective function is now the following:

$$\min TC = \sum_{u \in \mathcal{U}} \sum_{t \in \mathcal{T}} f_t \cdot (d_t \cdot (FC_{ut} + OC_{ut}) + SC_{ut}).$$

Besides the modification of the objective function, further modifications are required. To the best of the author's knowledge, these modifications are not considered in other models. These modifications concern all intertemporal restrictions. In the following, we demonstrate the need for modifications with simple examples, and propose two alternative formulations for the corresponding constraints.

Example (Assumptions). *For the following examples, we assume that every two months are represented by a weekday and a weekend day. Every day is represented by 12 timesteps. Let us further assume that the first day is a weekend day. Timesteps 1–12 are used to represent the first weekend day. Timesteps 13–24 are used for the first workday.*

Start-Ups and Shut-Downs:

Start-up restriction (2.7) and shut-down restriction (2.8) are used to determine the capacity started-up and the capacity shut-down at every timestep. These values are required for the minimum up-time and minimum down-time restrictions. Furthermore, the variable $P^{StartUp}$ is used to determine start-up costs. Inequalities (2.7) and (2.8) are intertemporal constraints, and need to be adjusted for a model based on typical days. In the following, we briefly demonstrate the need for a modified problem description with a simple example, before we propose two different formulations that take these problems into account.

---

[13]If every two months are represented by one workday and one weekend day, and there are 41 workdays in January and February, then f_t is 41 for each timestep representing a workday in this period.

[14]The duration of timesteps is not at all considered in [Sun+08].

Example (Strict start-up constraints). *Let us consider how the capacity started at time-step 13 is determined. Using start-up constraint (2.7), we get:*

$$P_{13}^{\text{StartUp}} \geq P_{13}^{\text{Online}} - P_{12}^{\text{Online}}. \tag{2.15}$$

This is correct, but only if we assume that the considered workday that is represented is a Monday. If the typical workday is not a Monday, the restriction would be the following:

$$P_{13}^{\text{StartUp}} \geq P_{13}^{\text{Online}} - P_{24}^{\text{Online}}. \tag{2.16}$$

As we assumed that the current day is not a Monday, the previous day is also a workday. Since we use just one workday to represent all workdays within two months, the previous day is represented by the same typical day as the current day. Hence, the previous timestep of timestep 13 is timestep 24 in this case. One possibility to cope with this problem is to use both constraints.

Formally written, we use in addition to the "normal" start-up constraint (2.7) the following restriction:

$$P_{ut}^{\text{StartUp}} \geq P_{ut}^{\text{Online}} - P_{u,t+t^{\text{PerDay}}-1}^{\text{Online}} \qquad \forall u, t \bmod t^{\text{PerDay}} = 1, \tag{2.17}$$

with t^{PerDay} being the number of timesteps per day. Note that this restriction is quite strict. Compared with the original problem (which is not based on representative timesteps), some feasible solutions are excluded. In the following, we refer to this formulation as *strict model*.

Example (Weighted start-up constraints). *Another possibility is to use a weighted sum of restrictions (2.15) and (2.16) to define P_{13}^{StartUp}. Therefore, we have to define two additional variables that are, as all variables we use, non-negative. Thereby, $P_{12,13}^{\text{StartUp}}$ denotes the capacity that is started from timestep 12 to 13, while $P_{24,13}^{\text{StartUp}}$ denotes the capacity started from timestep 24 to 13:*

$$P_{12,13}^{\text{StartUp}} \geq P_{13}^{\text{Online}} - P_{12}^{\text{Online}}$$

and

$$P_{24,13}^{\text{StartUp}} \geq P_{13}^{\text{Online}} - P_{24}^{\text{Online}}.$$

We can now define $P_{13}^{StartUp}$ as:

$$P_{13}^{StartUp} \geq \frac{1}{5} \cdot P_{12,13}^{StartUp} + \frac{4}{5} \cdot P_{24,13}^{StartUp}. \tag{2.18}$$

Note that the auxiliary variables $P_{12,13}^{StartUp}$ and $P_{24,13}^{StartUp}$ in inequality (2.18) cannot be replaced with the right hand sides of restrictions (2.15) and (2.16). The reason is that the right hand sides of the restrictions may get negative, while the variables are greater or equal to zero. A negative right hand side in one of both restrictions would offset the number of start-ups of the other restriction. Assume for example, $P_{12}^{Online} = 500$, $P_{13}^{Online} = 100$ and $P_{24}^{Online} = 0$. Without the additional variables, one would get $P_{13}^{StartUp} = 1/5 \cdot (100 - 500) + 4/5 \cdot (100 - 0) = 0$. As a consequence, no start-up costs are considered in this case, even though—considering a whole week—in total 400 MW of capacity are started.

To write the weighted start-up constraint (2.18) in a general form, we first have to define m_t as:[15]

$$m_t = \begin{cases} 5 & \text{if } t \text{ belongs to a workday,} \\ 2 & \text{otherwise.} \end{cases}$$

Thereby, m_t counts how often the day to which timestep t belongs is repeated, before the following typical day occurs. It is different from f_t defined earlier, which is a weighting factor that counts the total frequency of the hours represented by t, independent of the fact whether these timesteps occur in a row or with other timesteps in between.[16]

The general form of the weighted start-up constraint (2.18) is

$$P_{ut}^{StartUp} \geq \frac{1}{m_t} \cdot P_{u,t-1,t}^{StartUp} + \frac{m_t-1}{m_t} \cdot P_{u,t+t^{PerDay}-1,t}^{StartUp} \qquad \forall u, t \bmod t^{PerDay} = 1, \tag{2.19}$$

with

$$P_{u,t-1,t}^{StartUp} \geq P_{ut}^{Online} - P_{u,t-1}^{Online} \qquad \forall u, t \bmod t^{PerDay} = 1$$

and

$$P_{u,t+t^{PerDay}-1,t}^{StartUp} \geq P_{ut}^{Online} - P_{u,t+t^{PerDay}-1}^{Online} \qquad \forall u, t \bmod t^{PerDay} = 1.$$

---

[15]We assume here a normal week with five workdays and two weekend days. The definition can easily be adjusted to consider holidays.

[16]Assuming there are nine weekends in January and February, f_t for timesteps of a weekend day of this period would be 18, while m_t is 2.

We refer to this model as *weighted model*. The shut-down restrictions of the basic model can be extended the same way, either as described in the strict model or the weighted model.

Minimum Up- and Down-Time Constraints:

If just the minimum up-time constraint (2.11) and the minimum down-time constraint (2.12) from the basic deterministic model are used, a model based on typical days might deliver solutions that are infeasible in reality. This can be illustrated with the following simple example.

Example. *Assume that there is a unit with a minimum up-time of two timesteps. Assume further that the unit is started on the last timestep of a weekday, while on the last timestep of the previous weekend day, the unit was already started earlier. Then, on the first timestep of the weekday, we might turn off the unit. However, in four out of five cases (namely from Tuesday to Friday), this decision is not allowed in reality, as the unit was just started one timestep before.*

To handle this problem, the two different methods already used for the start-up and shut-down variables can be applied. We demonstrate how to use them on the example of the minimum up-time restriction. The first possibility, applied in the *strict model*, is to use in addition to the normal minimum up-time restriction (2.11) the following restriction:

$$P_{ut}^{\text{Online}} \geq \sum_{\tau=t-(t_u^{\text{MinUp}}-1)}^{t} P_{u,\hat{t}(\tau)}^{\text{StartUp}}$$

$$\forall u, \; \lceil (t - (t_u^{\text{MinUp}} - 1))/t^{\text{PerDay}} \rceil < \lceil t/t^{\text{PerDay}} \rceil, \quad (2.20)$$

with[17]

$$\hat{t}(\tau) = \tau + \left(\lceil t/t^{\text{PerDay}} \rceil - \lceil \tau/t^{\text{PerDay}} \rceil \right) \cdot t^{\text{PerDay}}.$$

Thereby, $\hat{t}(\tau)$ simply denotes the preceding timesteps of t that are represented by the same typical day. For example, if $t = 14$ and $\tau = 13$, then $\hat{t}(13) = 13$. If $\tau = 12$, then $\hat{t}(12) = 24$.

While the normal minimum up-time constraint (2.11) ensures that the minimum up-time constraint holds if the previous day is represented by another typical day

---

[17]We assume that the maximum minimum up-time and the maximum minimum down-time is not greater than 24 hours.

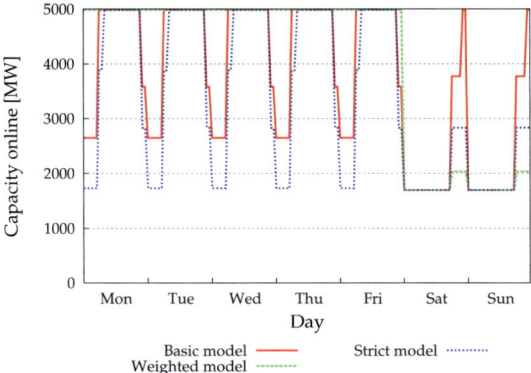

Figure 2.6: Capacity online of a group of coal units using different constraints.

(e.g., the current day is a Monday and the previous day was a Sunday), the additional minimum up-time constraint (2.20) ensures that the minimum up-time constraint also holds if the previous day is represented by the same typical day (e.g., Tuesday/Monday).

While this approach excludes some solutions that are feasible in the original problem, it guarantees that no solutions that violate the minimum up-time restrictions in reality are allowed in the model. In the *weighted model*, we combine the two minimum up-time restrictions (2.11) and (2.20) to

$$P_{ut}^{\text{Online}} \geq \frac{1}{m_t} \cdot \sum_{\tau=t-(t_u^{\text{MinUp}}-1)}^{t} P_{u,\tau}^{\text{StartUp}} + \frac{m_t-1}{m_t} \cdot \sum_{\tau=t-(t_u^{\text{MinUp}}-1)}^{t} P_{u,\hat{t}(\tau)}^{\text{StartUp}} \qquad \forall u, t.$$

(2.21)

If $\lceil (t-(t_u^{\text{MinUp}}-1))/t^{\text{PerDay}} \rceil = \lceil t/t^{\text{PerDay}} \rceil$, then the weighted minimum up-time constraint (2.21) corresponds to normal minimum up-time constraint (2.11). Note that the weighted formulation of the minimum up-time constraint allows infeasible solutions as well as it excludes feasible solutions, but both to a lower extend compared with the basic model and the strict model, respectively.

Figure 2.6 shows an example of the operation of a group of coal units for a week applying the basic model, the weighted model and the strict model. In this example, both the *weighted model* as well as the *strict model* lead to a feasible unit commitment. The unit commitment of the basic model is infeasible due to a violation of the minimum up-time constraint during the night of Saturday.

Storage Constraints:

It is very important to adjust the storage content constraint accordingly. The normal storage content constraint (2.13) ignores that timesteps have a different frequency. If just constraint (2.13) is used,[18] a very monotone storage operation as the one depicted in Figure 2.7(a) can be observed. During workdays, the storage produces electricity at its maximum level, while during weekends, the storage gets charged. Storage operation seems to be independent of electricity prices. At the moment of the highest prices, the storage is charged at the maximum charging level.

This behavior is caused by the non-consideration of the different frequencies of the timesteps. For the storage operation, the model assumes equal frequencies for all timesteps, which is not true. As there are more workdays than weekend days, much more electricity is produced compared to the electricity used to charge the storage. The obtained solution is infeasible in reality.

To get a more realistic storage operation, we use two additional constraints. For the last timestep of each representative day (i.e., $t + 1$ mod $t^{\text{PerDay}} = 0$), we replace the normal storage content restriction (2.13) with the following equation:

$$V_{u,t+t^{\text{PerDay}}} = V_{u,t} + \sum_{\tau=t+1}^{t+t^{\text{PerDay}}} (m_\tau \cdot d_\tau \cdot (\eta_u \cdot W_{u\tau} - P_{u\tau}))$$

$$\forall u \in \mathcal{U}^{\text{Storage}}, t \text{ mod } t^{\text{PerDay}} = 0. \quad (2.22)$$

Equation (2.22) ensures that the storage content at the end of a representative day corresponds to the the storage content at the end of the previous representative day adjusted by the electricity usage of the storage during the period represented by the representative day. For example, constraint (2.22) guarantees that the storage content at the end of Friday corresponds to the storage content at the end of Sunday adjusted by the production and charging during the whole week.

The second constraint we add ensures that the storage content at the end of the two months period represented by one weekday and one weekend day corresponds to the storage content at the last timestep of the previous period adjusted

---

[18]Note that also constraint (2.13) must be adapted using the duration of a timestep d_t.

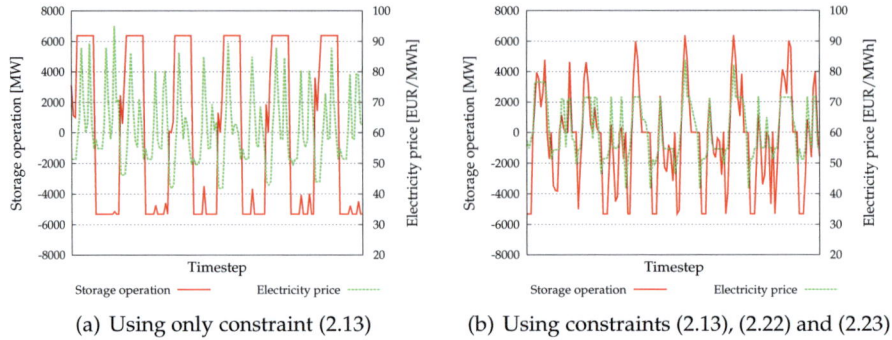

(a) Using only constraint (2.13)   (b) Using constraints (2.13), (2.22) and (2.23)

Figure 2.7: Storage operation (left axis) and electricity prices (right axis). Negative values for the storage operation correspond to storage charging, positive values represent electricity production.

by the storage production and charging within this period:

$$V_{u,t+2 \cdot t^{\text{PerDay}}} = V_{u,t} + \sum_{\tau=t+1}^{t+2 \cdot t^{\text{PerDay}}} \left(f_\tau \cdot d_\tau \cdot (\eta_u \cdot W_{u\tau} - P_{u\tau}) \right)$$

$$\forall u \in \mathcal{U}^{\text{Storage}}, t \bmod 2 \cdot t^{\text{PerDay}} = 0. \quad (2.23)$$

Examining Figure 2.7(b), it can be seen that considering constraints (2.22) and (2.23) in addition to constraint (2.13) gives a much more realistic storage operation. The storage operation is now strongly correlated with the electricity prices. During high price periods, electricity is produced, while during low price periods, the storage is charged. Compared with Figure 2.7(a), one can notice that the price peaks are now lower—a result of the more appropriate storage operation.[19] However, as the storage produces more electricity in the version depicted in Figure 2.7(a) than it can produce in reality, average electricity prices rise from 60.59 €/MWh to 62.49 €/MWh with the new constraints.

2.3.4   Stochastic Model Based on a Recombining Scenario Tree

As described before, an electricity market model with perfect foresight is not very realistic. However, owing to the path-dependence, a stochastic programming formulation of the model as described in Section 2.3.2 becomes computationally in-

---

[19]It should be remarked that the proposed formulation does not guarantee a feasible storage operation. However, compared to the standard formulation, the proposed changes are a significant improvement.

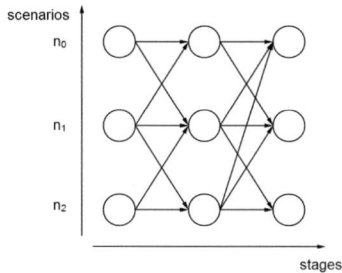

Figure 2.8: Illustration of a recombining scenario tree.

tractable if a longer time horizon is considered. There are two different methods that try to circumvent this problem.

The first possibility is to divide the problem into overlapping subproblems with a shorter time horizon. These subproblems can be solved as stochastic programming problem. This approach is called *rolling horizon* and is used for example by Meibom et al. [Mei+06]. However, using a rolling horizon approach, many smaller stochastic problems have to be solved. Depending on the level of detail, this can be very time consuming. Another problem of this approach is the management of hydro reservoirs. For a model considering a whole year it is reasonable to require the final reservoir content to be equal to the initial content. For a model with a short time horizon (e.g., a day or a week) this is not the case, as the water inflows exhibit seasonal variations.[20]

The second possibility to add stochasticity to an electricity market model is to relax the path-dependence of the problem and use a recombining scenario tree. Such a recombining scenario tree is depicted in Figure 2.8. This approach is followed in [SW07; Swi07; Sun+08; HS09]. The recombining scenario tree prevents the number of nodes to grow exponentially. However, the path-dependence of the problem cannot be considered. Therefore, average values are used for variables that depend on the values of preceding variables from different scenarios. Despite this disadvantage, we choose this approach as a rolling horizon model is too time consuming for our investment model. The way we adjust the model based on typical days to consider stochasticity is described in this section, especially the cases where our approach is different from the one applied by the other models in this area.

---

[20]A possible approach to deal with the problem is to determine a value for the electricity saved in the reservoir, which is added to the objective function. However, the determination of appropriate values for the saved electricity is a non-trivial task.

Analogous to the normal stochastic programming model, an index for the scenario is added to all variables that can take different values depending on the scenario. In our model, this are all variables except the ones used to model reserve energy. Similar to the case of the model based on typical days, problems arise through the time coupling constraints. The adaption of all other constraints is straightforward. As in the stochastic programming model, these non time-coupling constraints just have to hold additionally for every scenario.

Objective Function:

The objective function in this model is a combination of the objective function used in the stochastic programming model and the one used in the model based on typical days:

$$\min TC = \sum_{u \in \mathcal{U}} \sum_{t \in \mathcal{T}} \sum_{n \in \mathcal{N}} \pi_{tn} \cdot f_t \cdot (d_t \cdot (FC_{utn} + OC_{utn}) + SC_{utn}).$$

Start-Ups and Shut-Downs:

Weber et al. [Web+06] propose the following restrictions[21] to determine the capacity started:

$$P_{utn}^{\text{StartUp}} \geq \frac{1}{\sum_{n'} \psi_{t-1,n' \to t,n}} \cdot \sum_{n'} \left(\psi_{t-1,n' \to t,n} \cdot (P_{utn}^{\text{Online}} - P_{u,t-1,n'}^{\text{Online}}) \right) \qquad \forall u, t, n.$$

However, there are two issues with this formulation. First of all, the shut-down of units from one of the preceding nodes may offset the start-up of units from other preceding nodes. Furthermore, we think that it is not enough to consider just the transition probabilities for the weighting of the preceding nodes. In our opinion, it would be more appropriate to consider also the occurrence probability of the preceding nodes. We propose the following constraints, corresponding to the start-up constraints (2.7) of the basic model:

$$P_{utn}^{\text{StartUp}} \geq \frac{1}{\sum_{n'} (\psi_{t-1,n' \to t,n} \cdot \pi_{t-1,n'})} \cdot \sum_{n'} (\psi_{t-1,n' \to t,n} \cdot \pi_{t-1,n'} \cdot P_{u,t,n',n}^{\text{StartUp}}) \qquad \forall u, t, n,$$

(2.24)

---

[21]We adapted the formulation to our notation.

with $P_{u,t,n',n}^{\text{StartUp}}$ denoting the capacity that is started-up in timestep t and scenario n, when the previous scenario was n'. This capacity is defined as:

$$P_{u,t,n',n}^{\text{StartUp}} \geq P_{utn}^{\text{Online}} - P_{u,t-1,n'}^{\text{Online}} \qquad \forall u, t, n', n. \tag{2.25}$$

The extension to the model based on typical days is relatively straightforward. In the case of the *strict model*, we just have to apply the logic used in inequalities (2.24) and (2.25) to the strict start-up restriction (2.17).

For the *weighted model*, things are a bit more complicated, as we need additional variables. Applying the same idea to the weighted start-up restriction (2.19), we get:

$$P_{utn}^{\text{StartUp}} \geq \frac{1}{m_t} \cdot P_{u,t-1,t,n}^{\text{StartUp}} + \frac{m_t-1}{m_t} \cdot P_{u,t+t^{\text{PerDay}}-1,t,n}^{\text{StartUp}} \qquad \forall u, n, t \bmod t^{\text{PerDay}} = 1,$$

with $P_{u,t-1,t,n}^{\text{StartUp}}$ being the average amount of capacity started in t considering all possible preceding scenarios at the previous timestep $t - 1$:

$$P_{u,t-1,t,n}^{\text{StartUp}} \geq \frac{1}{\sum\limits_{n'}(\psi_{t-1,n'\to t,n} \cdot \pi_{t-1,n'})} \cdot \sum\limits_{n'}(\psi_{t-1,n'\to t,n} \cdot \pi_{t-1,n'} \cdot P_{u,t-1,t,n',n}^{\text{StartUp}})$$

$$\forall u, n, t \bmod t^{\text{PerDay}} = 1$$

and

$$P_{u,t-1,t,n',n}^{\text{StartUp}} \geq P_{utn}^{\text{Online}} - P_{u,t-1,n'}^{\text{Online}} \qquad \forall u, t, n', n.$$

The constraints for $P_{u,t+t^{\text{PerDay}}-1,t,n}^{\text{StartUp}}$ are defined accordingly.

Minimum Up- and Down-Time Constraints:

To be able to apply the minimum up- and down-time constraints (for either the *strict* or the *weighted model*), we define additional variables P_{ut}^{StartUp} that represent the average started capacity per timestep:

$$P_{ut}^{\text{StartUp}} = \sum\limits_{n}\left(\pi_{tn} \cdot P_{utn}^{\text{StartUp}}\right) \qquad \forall u, t.$$

Now, we simply have to add the index n to the minimum up-time and minimum down-time constraints of the chosen model.

Storage Constraints:

The storage restrictions have to be extended similar to the other time-coupling constraints. The "normal" storage content constraint (2.13) becomes:

$$
V_{utn} = \frac{1}{\sum_{n'}(\psi_{t-1,n'\to t,n} \cdot \pi_{t-1,n'})} \cdot \sum_{n'}(\psi_{t-1,n'\to t,n} \cdot \pi_{t-1,n'} \cdot V_{u,t-1,n'})
$$
$$
- (P_{utn} + \eta_u \cdot W_{utn}) \cdot d_t
$$
$$
\forall u \in \mathcal{U}^{\text{Storage}}, t \bmod t^{\text{PerDay}} \neq 0, n.
$$

For the adaption of equation (2.22), we use the following equation:[22]

$$
V_{u,t+t^{\text{PerDay}},n} = V_{utn} + \sum_{\tau=t+1}^{t+t^{\text{PerDay}}} \sum_{n'}(m_\tau \cdot d_\tau \cdot \pi_{t-1,n'} \cdot (\eta_u \cdot W_{u\tau n'} - P_{u\tau n'}))
$$
$$
\forall u \in \mathcal{U}^{\text{Storage}}, t \bmod t^{\text{PerDay}} = 0, n.
$$

Equation (2.23) is adjusted accordingly.

2.3.5   Price Determination

To calculate the contribution margin of power plants, the electricity production, the costs for this production and the electricity prices are required. While the determination of the first two parameters is obvious in an LP or an MILP electricity market model, there are different ways how electricity prices can derived. In the following, we briefly describe two of these methods, before we discuss a possibility to determine carbon prices.

Electricity Price Determination:

The probably most common method to determine electricity prices in fundamental market models described as an LP or an MILP problem is based on dual variables (for an introduction to duality in LP models see, e.g., Hillier and Lieberman [HL05]). The value of the dual variables of the power balance restriction (2.5) corresponds to the marginal costs of electricity production. Based on the assumption of perfect competition, the electricity prices should be equal to these marginal

---

[22]Instead of using the occurrence probability $\pi_{t-1,n'}$, one might use the conditional probability of each node between t and $t + t^{\text{PerDay}}$ under the assumption of being in node n at timestep t. For the sake of simplicity, we omit the calculation of the conditional probabilities and use $\pi_{t-1,n'}$ instead.

costs. This method to determine electricity prices is widely used (e.g., Burger et al. [Bur+07], Meibom et al. [Mei+06], Swider [Swi07]).

Note that this method is only directly applicable to LP models, as it requires *strong duality*, which is not guaranteed for MILP models. To derive shadow values for MILP models, all integer variables have to be fixed. The resulting LP–problem can then be used to get the shadow values. However, in such a case start-up costs are not included in the marginal costs. This issue is discussed by the author in [Bar+08].

Another approach to determine electricity prices, mentioned in [Bur+07, p. 178], is to vary the demand by $\pm\Delta d$ for a time period τ, and solve the problem for these two cases. Based on the cost difference ΔTC, specific marginal costs S'_τ can be derived as $S'_\tau = \Delta TC/(2\Delta d)$. The advantage of this method is that it considers start-up costs (and other integer constraints) also in case of an MILP. The disadvantage is that for each time period τ for which prices should be determined, two problems must be solved. To get hourly prices for a whole day, 48 problems must be solved. If electricity prices are determined based on dual variables, only one problem must be solved. Therefore, we use the approach based on dual values.

Carbon Price Determination:

Similar to the determination of electricity prices, the dual value of constraint (2.14) can be interpreted as carbon price. Using this method, the carbon price is derived based on fundamental factors such as the amount of emitted CO_2.

However, this method implies several simplifications compared to the real emission allowances trading system. First of all, only the electricity sector of Germany is considered for the price determination. No trades of emission allowances with other countries or other industry sectors are possible. Otherwise, Germany is by far the greatest emitter of CO_2 in the EU ETS and accounts for about 25% of the total emissions (European Commission [Eur07]). Thereby, about 70% of the emissions are allocated to the electricity sector (BMU [BMU07]).

Another simplification of this method is the non-consideration of banking (the transfer of non-used emission allowances from one year to another). At least within the different phases of the EU ETS, banking is possible.

Besides the simplifications discussed before, the proposed method is quite useful, as it determines carbon prices depending on the existing power plant fleet and fuel prices. Considering the difficulties of long-term carbon price forecasts, this method may contribute to obtain more realistic carbon prices. To integrate external influences on emission prices like the trade with foreign countries, the method can be combined with external price forecasts, using a weighted price of the external forecasts and the model endogenous price.

To be able to combine endogenous prices based on the value of the dual variables of constraint (2.14) with exogenous price forecasts, we solve our LP model twice. During the first iteration, no carbon costs are considered in the objective function. The costs of CO_2 used in equation (2.4) defining the fuel and carbon costs are set to zero. At the same time, the amount of allowed CO_2 emissions is restricted by constraint (2.14). After this problem has been solved, we determine the value of the dual variable of constraint (2.14). Then, we set the CO_2 costs equal to a weighted value of this endogenously determined CO_2 price and an exogenous CO_2 price forecast. We remove constraint (2.14) from the model, and solve the model again. Based on the results of this iteration, electricity prices, electricity production and the associated costs are determined.

2.4   Comparison with Observed Prices

For the investment models proposed in the following two chapters, any of the presented fundamental electricity market models can be used. We opt for the LDC model due to its low computational requirements. Besides the LDC model, we also use an LP-based model to evaluate the profitability of coal units with carbon capture and storage (CCS) technology.

In this section, we compare the electricity prices of these two models with observed prices. We start with a description of the input data that we use for the comparison. Then, we briefly discuss how model quality can be measured, before we compare the results of both models with EEX prices. The comparison is based on the year 2008 and the first half of 2009. Finally, we discuss the results of both models and motivate our choice for the fundamental electricity market model which we choose for the investment model in the following chapters.

2.4.1   Input Data

In this subsection, we describe the input data that we use for our fundamental models. We consider the German electricity market. Foreign markets are not explicitly modeled, but included with time-series of expected cross-border electricity exchanges. For model calibration, data from the years 2006 and 2007 is

	Unit	Nuclear	Lignite	Coal	Gas CC	Gas GT	Oil
Availability	%	94.0	84.7	82.1	88.7	90.7	90.7
Capacity	GW	20.3	21.5	26.2	14.6	4.6	4.4
Minimum Up-Time	h	12	8	8	8	0	0
Minimum Down-Time	h	8	8	8	4	0	0
Operating costs	€/MWh	0.5	1.6	2.0	1.2	1.2	1.2
Start-up costs	€/MW	2.0	5.0	5.0	8.0	8.0	8.0
Start-up fuel usage	MW_{th}/MW_{el}	16.7	6.2	6.2	3.5	1.1	1.1

Table 2.2: Technical parameters of the considered units. Sources: Bagemihl [Bag02], Swider and Weber [SW07], Swider et al. [Swi+07].

used. In the following, we describe how electricity supply and electricity demand are modeled, and we give an overview of the fuel and carbon prices used for the comparison.

Electricity Supply:

The power plants considered in our model correspond to the existing units in Germany. The data is combined from different publicly available sources like the EEX, data provided by generation companies, and a list of power plants of the German Federal Environmental Agency [Umw09]. We adjusted the data by adding a coal unit, an oil-fired unit and a lignite unit such that the total capacity of each fuel type corresponds approximately to the capacities provided by the German Federal Statistical Office [Des08].

For each unit, we multiply the capacity with an availability factor. This factor is used to account for non-availabilities of plants due to maintenance or forced outages. The values for the availabilities are taken from Swider et al. [Swi+07, p. 67]. In the LDC model, we are constrained to use a constant factor for each unit for the whole year. As most planned maintenance is scheduled in the summer months when demand is low, a seasonable availability factor is more realistic than a constant one. In the LP model, we consider such seasonal availabilities.

Table 2.2 shows the constant availability factors used in the LDC model as well as other parameters of the considered units. As far as the efficiencies of the plants are publicly available, we use these efficiencies. For those plants where the efficiencies are unknown, we approximate the efficiency based on the plant type, the fuel and the age of the plant, as described by Schröter [Sch04].

Electricity Demand:

For the electricity demand, hourly load profiles provided by the European network of transmission system operators for electricity (ENTSO-E) are used.[23]

As there is an obligation to feed-in electricity produced by renewable sources, we directly subtracted the production of hydro-units (run-of-river plants and hydro reservoirs with a natural inflow) and wind power units from the demand. For the calculation of the (expected) hydro-production, we use the time-series of hydro-production from the publicly available database of the EU project *Wind Power Integration in Liberalised Electricity Markets (WILMAR)*. [24] Concerning the electricity produced by wind power, we use two different data sources. For the electricity produced by onshore wind units, we use the (hourly) wind power feed-in provided by the four TSOs (EnBW, Amprion, transpower, 50Hertz). As the offshore wind speed characteristics are different from the onshore ones, we use the (offshore) wind speed time-series of the *FINO (Forschungsplattformen in Nord-und Ostsee)* project[25] to derive the expected offshore wind production. The wind speed forecasts are transformed to power using a power transform curve (see, e.g., McLean [McL08]).

The increased amount of wind power leads to an increasing demand of control power. Therefore, we model the control power demand depending on the installed wind power capacity as described by DEWI et al. [DEW+05]. In the LDC model, we add the control power demand to the total electricity demand, as proposed by Burger et al. [Bur+07, p. 174].

Additionally, electricity demand is adjusted by electricity exchanges with neighboring countries. This is based on hourly time-series of cross-border physical load flows published by the TSOs.[26]

---

[23]These hourly load profiles are available in the "Country Package - Production, Consumption, Exchange" at http://www.entsoe.eu/, last time accessed December 01, 2009

[24]For a documentation, see Kiviluoma and Meibom [KM06]. The WILMAR database is available at http://www.wilmar.risoe.dk/Results.htm, last time accessed December 01, 2009.

[25]Data provided by the *Bundesministerium für Umwelt, Naturschutz und Reaktorsicherheit (BMU)*, the *Projektträger Jülich (PTJ)* and the *Deutsches Windenergie Institut (DEWI)*. The data is available (after registration) at http://www.bsh.de/de/Meeresdaten/Beobachtungen/Projekte/FINO/, last time accessed December 1, 2009

[26]As there are no transmission values to Sweden published by transpower, we took this data from NordPool.

	Coal	Gas	Carbon
2006	2.11 €/GJ	5.93 €/GJ	17.40 €/t
2007	2.33 €/GJ	5.55 €/GJ	0.68 €/t
2008	3.77 €/GJ	7.45 €/GJ	22.45 €/t
2009[27]	2.35 €/GJ	3.80 €/GJ	12.68 €/t

Table 2.3: Fuel and carbon prices. Sources: BMWi [BMW09], EEX.

Fuel Prices:

The fuel and carbon prices that we use for the comparison are shown in Table 2.3. While electricity demand and wind power production are based on the values of 2006 and 2007, we use the observed fuel prices of 2008 and 2009 for the following comparison. This choice is motivated by the objective to test how well our models perform under the assumption of pertinent fuel prices. For the power generation expansion problem, fuel price uncertainty is reflected by considering different fuel price scenarios.

2.4.2 Measuring Model Quality

An important issue related to modeling in general is the question how to measure model quality. For descriptive models, different measures comparing the predicted values with the observed values in reality are proposed. Examples for such measures are the goodness-of-fit measure, the root mean squared error (RMSE), the mean absolute error (MAE), the mean percentage error and the value of the likelihood function (Weber [Web05, p. 27]). The appropriate measure depends on the use of the model and the predicted values.

For the assessment of the quality of electricity prices predicted by our fundamental electricity market models, we use the MAE and the RMSE. This choice is motivated by the fact that for price forecasts measures of absolute errors are more meaningful than measures of relative errors [Web05].

The RMSE is defined as

$$\text{RMSE} = \frac{1}{n} \sqrt{\sum_{i=1}^{n} (x_i - \hat{x}_i)^2},$$

---

[27]Data for the year 2009 is based on the period between January and July.

while the MAE is defined as

$$\text{MAE} = \frac{1}{n} \sum_{i=1}^{n} |x_i - \hat{x}_i|,$$

with x_i being the prices observed in reality, $\hat{x}_i$ the prices predicted by the model and n the number of observations.

Concerning the data sets that are used to assess the model quality, there are two different options. The first possibility is to evaluate a model with the data sets that were used to calibrate the model. However, this is only a weak indicator of model quality, as the model generally delivers the best results for these data sets. Nevertheless, such a comparison may be performed to get a first idea about model quality especially if the purpose of the model is not predicting future values but, for example, comparing different market designs.

A more appropriate indicator of model quality is to test the models with data sets that were not used for calibration, but for which data samples are already observed. A common approach is to use data sets for model calibration that are two to three years old. Model quality can then be assessed by comparing the predicted values for the following year with the observed values. We follow this approach and use the years 2006 and 2007 for model calibration, while we compare the model results of 2008 and the first half of 2009 with observed EEX prices.

2.4.3   Load–Duration Curve Models

In this subsection we compare the electricity prices obtained with our LDC model with observed EEX spot prices. We use three different versions of the LDC model: the basic model without adjustments, a model using the linear adjustments of the function (2.1), and a model using in addition the scarcity mark-ups described by function (2.2). We refer to these models as *LDC*, *LDC linear*, and *LDC SMU*, respectively.

Table 2.4 shows average, minimum and maximum prices, the standard deviation, the MAE, and the RMSE for the different models for the years 2008 and 2009.

All LDC models approximate average electricity prices quite good. However, the volatility of model prices is much lower than the volatility of EEX prices. The reason for the lower volatility is the non-consideration of technical restrictions and

---

[28]The lower average price for the *LDC linear* model compared to the *LDC* model is due to the consideration of the period between January and July. Considering the whole year, the average electricity price would be equal for both models.

	EEX	LDC	LDC Linear	LDC SMU
		2008		
Average price	65.76	63.99	63.99	65.42
Min. price	−101.52	42.34	22.47	22.47
Max. price	494.26	132.31	152.87	171.68
St. dev.	28.66	7.15	14.74	17.23
MAE	–	16.02	9.05	8.08
RMSE	–	21.75	14.66	12.19
		2009		
Average price	38.43	36.93	36.41[28]	37.22
Min. price	−151.67	31.15	11.31	11.31
Max. price	159.92	44.68	61.27	74.23
St. dev.	17.22	2.49	10.19	11.57
MAE	–	9.69	3.84	3.01
RMSE	–	15.05	9.03	7.24

Table 2.4: Model results for 2008 and January–July 2009. All values are given in €/MWh. MAE and RMSE are calculated based on sorted prices.

uncertainty. Compared with the other two models, the LDC model without adjustments produces the highest MAE and RMSE. As expected, the two other models produce a higher volatility, even though the volatility of both models is still considerably lower than the volatility of EEX prices. Concerning average prices, the MAE as well as the RMSE, the model with scarcity mark-ups delivers the best results for the considered periods.

Similar conclusions can be derived by looking at the price–duration curves shown in Figure 2.9. Note that the model with scarcity mark-ups and the model using only linear adjustments deliver the same prices during periods of low load (therefore, the *LDC linear* curve is hidden behind the *LDC SMU* curve during these periods).

Electricity prices are better approximated for the first half of 2009 than in 2008. It is remarkable that the lowest electricity prices are overestimated by all models. The fact that the model with mark-ups delivers the best results of the three models is not surprising. An explanation for this is the structure of the merit-order curve, which is non-linear. Deviations in the case of scarcity have the highest impact, as the merit-order curve is the steepest in this region. This characteristic can be better approximated by a piecewise-linear function than by a linear one.

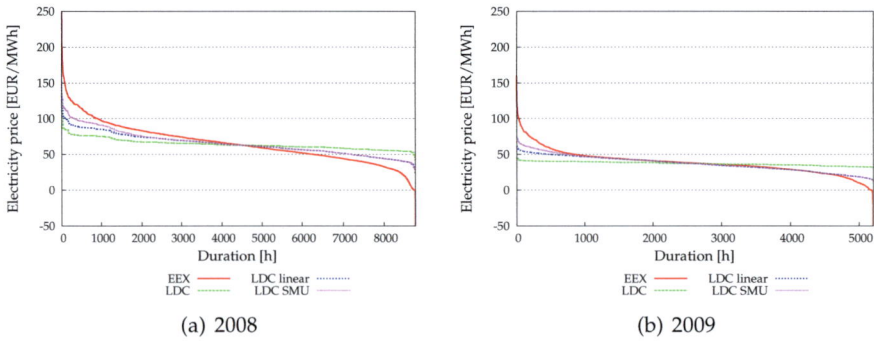

Figure 2.9: Price–duration curves for 2008 (a) and January - July 2009 (b).

The RMSE and the MAE error are, especially in 2008, relatively high. The consequences of a high RMSE and MAE of electricity prices used in an investment model depend on the types of units considered. If peak load units, which operate only during few hours a year, should be evaluated, then high RMSE and MAE of electricity prices can result in significant errors concerning the contribution margins of these units. If base load units, which operate during most hours of a year, are evaluated, average electricity prices are the more important quality measure than the RMSE and the MAE. Assuming that the CCGT units and coal units we evaluate are at least mid load units, we argue that the quality of the prices obtained with our LDC models is sufficient for the use in our investment models.

2.4.4 Linear Programming Models

In this subsection, we present the results of different LP models presented in Section 2.3. We test deterministic and stochastic models and compare the results of three different model formulations: the strict model, the weighted model, and the model without the additional constraints used for the two other models. We refer to this model (which corresponds approximately to the models used in literature) as basic model in this subsection. Note that it does not correspond to the basic deterministic model presented in Section 2.3.1, but is based, as the other considered models, on typical days.

Besides the representation of a complete year based on typical days, another simplification is used to reduce the computational burden: units of the same technology and similar efficiencies are aggregated to unit groups. Before we present the model results, we describe how the load-profiles and the wind power scenarios for the typical days are selected.

Determination of Typical Load Profiles:

For the comparison presented in this section, we use six typical weekdays and six weekend-days, each representing a period of two months. Each day consists of twelve timesteps of two hours. To determine typical load profiles for these timesteps, we use the relative hourly load profiles[29] of 2006 and 2007.

For every typical day, we transform the 24 hours load-profiles to 12 timesteps load-profiles by taking the average value of every two hours. We divide these load profiles into different clusters based on the timestep of the day, the weekday (weekend/workday) and the month.

To select a representative load profile for each of these clusters, different methods can be applied:

- The simplest one is to use average values for every timestep of the day. However, using average values results in a non-consideration of extreme scenarios. Especially the ignorance of high-load scenarios is critical, as the merit–order curve is very steep in this region. Disproportionately high prices which occur in high-load periods cannot be captured by a model based on average values.

- To circumvent this problem, one might consider additionally high-load (and low-load) scenarios. Such scenarios can be chosen for example using the *fast forward selection*, a method commonly used in stochastic programming to construct scenarios trees (see, e.g., Gröwe-Kuska et al. [GK+03]).[30]

  While the inclusion of high-load and low-load scenarios led to slightly more accurate results during empirical tests we did, it has the disadvantage to increase the problem size by a factor of three.

- An alternative method to better represent the LDC is the following: After the selection of the representative load scenarios for every typical day (which

---

[29]The hourly load relative to the total annual load.

[30]The *fast forward selection* is a heuristic for the minimization of the distance between a predefined number of selected scenarios and all other scenarios. The algorithm iteratively selects a scenario and adds it to the set of selected scenarios, until the predefined number of selected scenarios is reached. To select a scenario, the algorithm calculates for every scenario s the sum of the minimum distance of all other scenarios to one of the already selected scenarios or s. The scenario s^* with the minimum sum of distance is added to the selected scenarios.

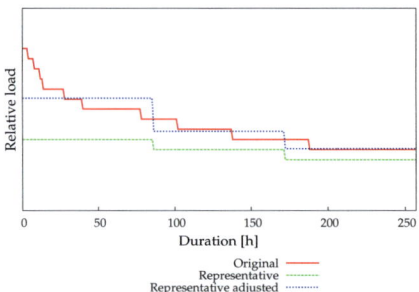

Figure 2.10: Original, representative and adjusted representative LDCs.

can be based on average values for example), these scenarios are used to build the representative LDC. Every timestep of the representative LDC is adjusted such that it equals the average value of the corresponding hours of the original LDC for the whole year.

This procedure is depicted in Figure 2.10 for periods of maximum load. The red curve is the original LDC based on all hourly load values. The green line represents the chosen representative (non-adjusted) load profiles. It can be seen that the non-adjusted representative LDC is significantly below the original LDC in this section. Therefore, the representative LDC is adjusted as described before. The result is the blue adjusted LDC curve in the figure.

We opt for the last method, as it allows a better approximation of the LDC compared to the first method, while it does not increase the computational complexity of the LP model.

Determination of Wind Power Scenarios:

The selection of wind power scenarios is based on hourly time series of the wind power capacity factor (the production divided by the capacity). As the offshore wind speed characteristics are different from the onshore ones, we use two different time series.[31] For onshore wind production, we derive the time series based on the hourly wind feed-in published by the TSOs (as before, the base for the model calibration are the years 2006 and 2007). The time series for offshore wind production is based on wind speed measurements of the FINO project.

---

[31]For the comparison done in this section, we use only the existing onshore capacity. Nevertheless, we include the offshore capacity in this description here, to illustrate how representative wind scenarios for future years used in the investment model are derived.

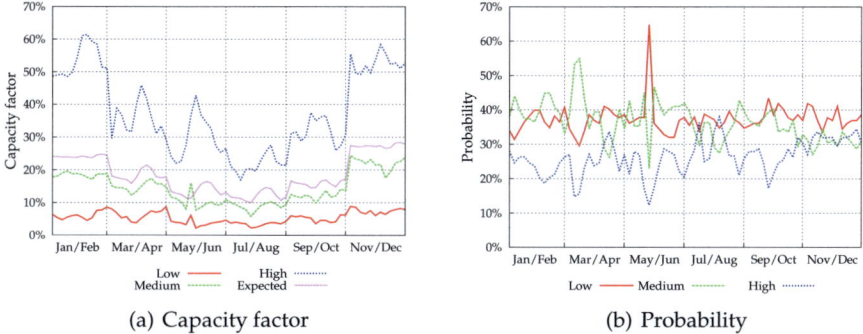

(a) Capacity factor   (b) Probability

Figure 2.11: Capacity factor and corresponding probability for the representative scenarios.

These two time series of the capacity factor are used for all years within the optimization horizon. As one timestep in our model corresponds to two hours, we calculate the average capacity factor for every two hours for offshore and onshore wind power. To determine the wind power scenarios for a single year, we multiply both time series with the installed onshore and offshore capacity, respectively. We combine these two time series and obtain a single time series with the expected average wind power production for every two hours of the year.

This time series is divided into several clusters. As for the determination of the average electricity demand, we consider periods of two months. For each of these periods, we determine the wind power scenarios for each of the twelve daily timesteps. Note that for the wind power scenarios, it makes no sense to distinguish between weekdays and weekends. For each of these typical timesteps, we have now about 120 different wind scenarios (as the basis for the scenario creation are two years, we are considering four months, e.g., January and February 2006 and 2007). To select the representative wind power scenarios, we use the *fast forward selection*.

The occurrence probability for each (reduced) scenario *s* is determined based on the number of non-representative scenarios that are closest to *s* (compared to the other representative scenarios). The transition probabilities between the different timesteps are calculated based on the transitions of all scenarios (the representative as well as the non-representative scenarios).

Figure 2.11 shows the resulting reduced scenarios and the corresponding probabilities for the year 2009. In Figure 2.11(a), we additionally included the expected capacity factor. It can be seen that the capacity factor is the highest in the period between November and February. Furthermore, the capacity is higher during the day than during the night.

It is interesting to see that we get quite different representative scenarios compared with Swider and Weber [SW07], who did a scenario selection based on cluster analysis. While our medium scenario is always except once above the expected capacity factor, the one in [SW07] is always below it. While we get approximately the same probabilities for the medium and low wind scenario and slightly lower probabilities for the high wind scenario, the low wind scenario dominates with a probability of almost 60% in [SW07].

The choice of only three scenarios has been made to reduce the computational burden. However, one problem specific to the choice of a very small number of scenarios and the model based on typical days should be mentioned. Due to the consideration of wind power scenarios from different days used for the same typical timestep, these scenarios are quite different, as it can be seen in Figure 2.11(a). As a result, the transition probability between the different timesteps and scenarios, e.g., from the low wind scenario to the medium wind scenario are relatively low. In most cases, only the "neighboring" scenario can be reached. Sometimes, the same succeeding scenario is selected (e.g., from low wind to low wind) with probability 1.

Model Results:

Table 2.5 shows the results obtained by different models for 2008 and January–July 2009. As the results of the different deterministic models are similar, only one of the three deterministic models, the weighted model, is shown. The standard deviation of all model prices is significantly lower than the one observed at the EEX, and average prices are about 3 €/MWh to 4.50 €/MWh lower than observed prices. Besides the general problem of fundamental models (e.g., data inaccuracy), these deviations are probably also caused the applied simplifications, especially the simplified representation of uncertainty and the selection of typical days.

Compared with the stochastic models, the deterministic models perform worse. Standard deviation is smaller, while the MAE and the RMSE are greater. Comparing the three stochastic models, it can be seen that the basic model delivers—as it can be expected—the lowest average prices. The strict model delivers in 2008 the most appropriate results except for the average price, which is best approximated by the weighted model. In 2009, the weighted model delivers the best results for all examined criteria. While the basic stochastic model performs worse compared

	EEX	Det. Weighted	Stoch. Basic	Stoch. Weighted	Stoch. Strict
		2008			
Average price	65.76	61.06	61.02	62.22	62.13
Min. price	−101.52	37.25	31.01	31.01	31.01
Max. price	494.26	107.65	162.53	169.53	172.87
St. dev.	28.66	14.52	17.20	17.48	17.57
MAE	−	12.39	11.01	10.84	10.80
RMSE	−	16.48	14.14	13.72	13.62
		2009			
Average price	39.12	35.68	35.62	36.05	35.88
Min. price	−151.92	30.71	21.65	24.18	24.22
Max. price	159.92	58.27	59.51	60.83	60.83
St. dev.	18.00	6.33	6.81	6.93	6.87
MAE	−	9.58	8.95	8.90	8.98
RMSE	−	13.47	12.50	12.39	12.49

Table 2.5: Model results for 2008 and January–July 2009. All values are given in €/MWh. MAE and RMSE are calculated based on sorted prices.

with the weighted and the strict stochastic models, the differences are less distinct compared to those between stochastic and deterministic models.

2.4.5 Conclusion

Compared with LDC models, the advantage of LP models is their ability to consider technical constraints, pumped-storage operation and uncertainties. However, due to the increased computational burden, simplifications must be made. These simplifications are one reason for deviations between model prices and EEX prices. The consideration of some typical days instead of all hours leads to longer horizontal sections on the LDC and therefore also to longer horizontal sections on the price–duration curve. The aggregation of units with the same technical constraints and similar efficiencies has a comparable effect. Several units with slightly different variable costs are represented by one unit with variable costs corresponding to the weighted average of the single units.

Figure 2.12 shows the price–duration curves for EEX prices, the stochastic, weighted LP model, the LDC model and the LDC model with scarcity mark-ups for the

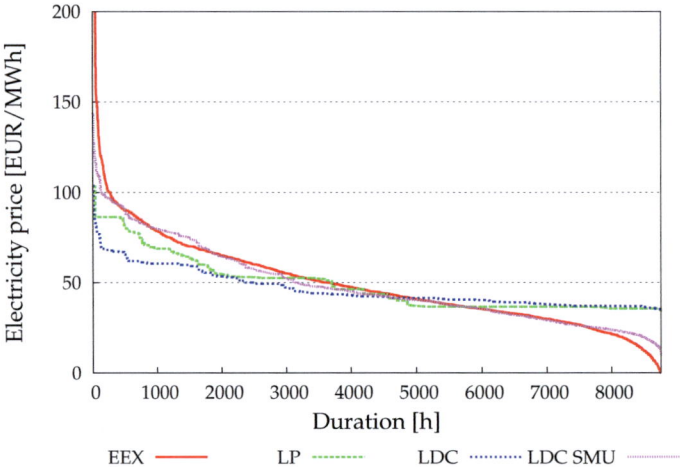

Figure 2.12: Price–duration curves 2006.

year 2006.[32] Comparing the curves of the LDC model with mark-ups and the LP model, it can be seen that the LDC curve is smoother. While the LP model approximates the price–duration curve better than the LDC model without adjustments, the LDC model with scarcity mark-ups performs best. This result can also be seen by comparing the results of the years 2008 and 2009 shown in Table 2.4 and Table 2.5.

Even though the LDC model with scarcity mark-ups performs better and is faster than the LP model, there are reasons for using the LP model nevertheless. First of all, the mark-up function used for the LDC model has been derived based on data of the years 2006 and 2007. While data of 2008 and 2009 was not used for the function fitting, it is probable that the structure of the electricity market in these years is similar to the one during the periods used for model calibration. Even though the adjustment function has been chosen with great care to avoid problems such as negative contribution margins, it is not clear whether the mark-up function we use is still pertinent if the structure of the market changes.

Second, using an LP model, carbon prices can be determined endogenously. This reduces one source of uncertainty, and is—due to the complexity of modeling carbon prices—an important feature. Therefore, we use the LP model in the case study presented in Chapter 5.5 focusing on the profitability of coal units with the

---

[32]We depict the values of 2006, as these values demonstrate the differences between the models the most clearly. The curves for the other years are similar, but the differences are less distinct.

CCS technology. As the relative long running times of the LP model make an extensive evaluation of model results impossible, we use for all other case studies in Chapter 5 the LDC SMU model.

2.5 Summary

In this chapter, we presented three different types of fundamental electricity market models, which can be used to determine the contribution margins of new units for an investment model. These three types of models are the LDC model, the SDC model, and models based on an MILP or an LP formulation of the electricity market.

The advantage of the first two types of models is their low computational burden compared to LP and MILP models. However, neither of them can consider technical restrictions of power plants, nor can they consider uncertainty. The difference between LDC models and SDC models is that LDC models cannot consider price-elastic demand while SDC models can. In addition, LDC models are based on the assumption of a constant merit-order curve for the considered period, while SDC models can use different merit-order curves. As LDC models require less computations compared to SDC models, we focused more on LDC models in this chapter.

The effects of the non-consideration of technical restrictions, uncertainty and the assumption of a constant merit-order curve on electricity prices have been discussed in detail. In general, the model tends to underestimate prices in times of high load periods, while it overestimates prices during low load periods. To account for these effects, we have proposed two different functions to adjust model prices. The first one should simulate the effect of start-up costs and avoided start-up costs as well as minim down-time restrictions and minimum up-time restrictions. Without changing average prices, we apply linear adjustments depending on the load relative to the average load. If load is above average load, prices are increased, while in case of load below average, prices are decreased. The second function we use is based on the scarcity of supply. In times of scarce supply, markups are added to the model's prices.

The advantage of electricity market models based on an MILP or an LP formulation is their ability to consider technical restrictions of plants as well as uncertainty. In addition, they can be used to determine the carbon price endogenously. However, due to the detailed description of the electricity market, the computational burden of these models is much higher than the one of LDC models and SDC models. To be able to use an MILP or an LP model for the determination of contribution margins in an investment model, several simplifications must be made. The exact MILP formulation is often replaced with an LP approximation, which can be solved faster in general. In addition, units with similar technical restrictions are often aggregated to unit groups. Furthermore, instead of considering a whole year, some typical days are selected for the optimization.

These simplifications make the original MILP formulation inapplicable, a fact that is often disregarded in literature. The relaxation of the MILP formulation, and also the aggregation of units to unit groups, requires an adjustment of the minimum up-time and minimum down-time constraints. We have proposed a formulation, which is applicable in these cases, and which is also more efficient than the formulation commonly used in literature.

Diverse adjustments—to the best of the author's knowledge not yet discussed in literature—that are required due to the use of typical days have been made. These adjustments concern all intertemporal constraints like start-up constraints, minimum up-time constraints, and minimum down-time constraints. It was shown that it is especially important to adjust the storage constraints to account for different frequencies of the representative timesteps. As a result, a much more realistic storage operation is obtained.

The selection of a few representative days has a further drawback: Depending on how these representative days are selected, they do not consider extreme scenarios, e.g., days with extreme high load. Due to the non-linearity of the merit-order curve, prices increase disproportionately strong in case of high load. Disregarding extreme scenarios results in an underestimation of prices. We have proposed a method to adjust the selected representative load scenarios such that the original load–duration curve is better approximated.

Besides the deterministic models, an approach based on recombining scenario trees to consider short-term uncertainties like the feed-in from wind power was presented. Even though this approach is a simplification of the computationally intractable stochastic programming formulation based on scenario trees, it delivers better results than the deterministic models.

In the last section of this chapter, we have compared LDC model prices and LP model prices with EEX spot prices in 2008 and the first seven months of 2009. It was shown that the proposed adjustments to the LDC models result in a better approximation of EEX prices compared to the standard LDC model. While average prices are well approximated by the LDC models, the MAE and MRSE are, especially in 2008, relatively great. However, as the units we consider in our investment model are at least mid load units, we argue that a good approximation of average electricity prices is more important than the MAE and the RMSE. Hence, we conclude that the LDC model is suited as a basis for our investment models.

The comparison of electricity prices obtained by the LDC model with the proposed adjustments with electricity prices obtained by the LP models delivered an unexpected result: Despite the drawbacks of the LDC model, it approximated observed spot prices better than the LP models. The reason for this result are probably the simplifications that are required to reduce the running time of the LP models. Despite these results, LP models are useful for two reasons: first, as LP models consider technical constraints, they are better suited to adapt to changing markets, and second, they allow to determine carbon prices endogenously.

For the investment models described in the next two chapters, any of the presented fundamental electricity market models can be used. Due to its fast running times and its good approximation of average prices, we use the LDC model with the proposed adjustments for most of the case studies presented in Chapter 5. For the case study examining the profitability of CCS units, we opt for the LP model due to its ability to determine carbon prices endogenously.

Chapter 3

Investment Model Based on Stochastic Dynamic Programming

In this chapter, we present an SDP formulation of the power generation expansion problem. Thereby, we model the coal price, the gas price and the carbon price as stochastic parameters. The contribution margins of the states are determined by one of the fundamental electricity market models presented in the previous chapter.

As a straightforward implementation of a normal SDP formulation results in a computationally intractable problem, we have to apply different techniques to simplify the problem. A common technique to reduce the complexity of an SDP problem is aggregation. We use two different forms of aggregation. The first one is temporal aggregation, a technique which is often used in SDP. Instead of considering stages of one year, we aggregate four years into a stage. This aggregation has the advantage that we can consider construction times of new units without adding extra attributes to a state.

The second form of aggregation is problem specific, and—to the best of the author's knowledge—not yet used in literature. Instead of explicitly considering coal, gas and carbon prices, we use the climate spread, which represents the cost difference between coal and CCGT units.

The model we propose differs in two points from similar SDP based formulations of the power plant investment problem (e.g., Jensen and Meibom [JM08], Klein et al. [Kle+08]). First, we consider investments by competitors. For this purpose, we propose two different methods to model such investments. One approach is based on the scarcity of supply, while the other one reflects in addition the expected profitability of new units.

Secondly, to represent fuel and carbon price uncertainty in an appropriate way, we use more fuel and carbon price scenarios than similar models in literature. To be able to handle a great number of different fuel and carbon price scenarios, we introduce another approach not yet used in this area: adaptive grid methods. Instead of solving the original problem, we solve an approximated problem consisting of a subset of the state space. This approximated problem is obtained by placing a grid in the state space. All grid points correspond to the states that are used for the approximated problem. To obtain a good solution quality, we iteratively solve the approximated problem, and then adaptively refine it by adding some new states.

The refinements we do are based on priorities calculated by a splitting criterion. We distinguish two kinds of splitting criteria: value-function based criteria and policy based criteria. While the first ones try to minimize the approximation error, the latter ones refine the problem around the optimal policy. We present different value function based splitting criteria, a policy based one and a combination of both. It turns out that the combination of both approaches works best in our problem setting.

While we consider only a subset of the state space, we do not restrict the decision space. Consequently, we may choose a decision that leads to a non-considered state. Hence, we need an approximation of the value of non-considered states. We obtain these approximations by interpolation. For this reason, we refer to our approach as interpolation-based stochastic dynamic programming (ISDP). We present and compare two different interpolation methods: multilinear interpolation and simplex-based interpolation. Multilinear interpolation uses more states for the interpolation than simplex-based interpolation. Hence, multilinear interpolation is slower, but yields generally more accurate results. Numerical tests show that for our problem, the more accurate interpolation results outweigh the slower running time.

Compared to a normal SDP model, our ISDP approach significantly reduces the number of considered states. As a consequence, uncertainties can be modeled in more detail. To be able to compare the performance of both approaches, we have to use relatively small problems that can also be solved by the normal SDP approach. Even for these small problems, the ISDP approach is preferable under certain conditions.

The remainder of this chapter is organized as follows: We start with a general description of the different parts of an SDP problem, namely the states, the decisions and the transitions. Next, we explain how we derive the fuel and carbon price scenarios for our model and how they are used to calculate the contribution margins for the states. Then we describe how we model investment decisions of competitors, before we present in detail the adaptive grid method. Afterwards, we give an overview of the different steps of the algorithm that we use to solve the problem, and comment on some design choices and implementation details. Finally, we evaluate our algorithm and conclude the chapter with a brief summary.

3.1 Problem Representation

In this section, we give a formal description of a straightforward implementation of the power plant investment problem as an SDP model. Thereby, we model fuel prices and carbon prices as stochastic parameters. We discuss the representation for the states, the decisions and the transitions in the SDP problem, and we outline how the rewards associated to the states and the costs of the decisions are determined.

3.1.1 States

An important design decision for the power generation expansion problem is the choice of an appropriate representation for the dynamic programming formulation. The chosen representation should catch all important characteristics of the model without being too detailed. An optimal representation rarely exists for real-world problems. In general, most problems are too complex to be modeled in full detail.

In the power generation expansion problem, the representation of a state should be chosen such that the contribution margin of this state can be determined with as much accuracy as possible. The appropriate description of a state depends on the way electricity prices, fuel costs and electricity production are modeled. As we use a fundamental electricity market model to derive these values, an ideal state representation should contain all relevant input parameters for such a fundamental model.

The difficulty is that there are many relevant input parameters: electricity demand, fuel prices, carbon prices, and the power generation fleet. These input parameters can be grouped into two parts: parameters that depend on the decisions in the dynamic problem, and parameters that do not depend on these decisions.

Obviously, parameters of the first group must be modeled as state attributes. In our problem, the new units belong to this group. Parameters belonging to the second group only have to be explicitly modeled, if different scenarios for these parameters should be considered, i.e., if they should be modeled as stochastic parameters. Because of their high volatility and their important influence on the contribution margins of new plants, we model the fuel prices and the carbon price as stochastic parameters. In this subsection, we use a simple example to derive the state representation that we use in our model. Thereby, we start with a deterministic problem, ignoring the construction times of new units. Next, we add the construction times to the model, before introducing stochastic fuel prices. Finally, we add investments of competitors to the problem. As this model is too complex to be solved, we introduce temporal aggregation to reduce the state space of the problem. Finally, we outline how the profits associated to a state are determined.

Deterministic State Representation Without Construction Lead Time

We start with a simple model, making several simplifications. First, we assume that all relevant data is known with certainty. This assumption implies that fuel prices, carbon prices and electricity demand are known as well as the investment decisions of competing companies. Second, we assume that new units are available in the year after the investment decision is taken.

In such a simple model, the states of a stage differ only in the number of new units. Therefore, it is sufficient to use the number of new units of each type as state variables. A graphical representation of this problem allowing investments in one CCGT unit and one coal unit is depicted in Figure 3.1.

For such a simple representation, the state space is manageable. Let $\mathcal{U}$ be the set of all investment alternatives, e.g., $\mathcal{U} = \{$CCGT unit, coal unit$\}$. Elements of $\mathcal{U}$ are denoted with u. The number of new units of type u at time t is x_t^u. Let $x^{\max,u}$ be the maximum number of new units of unit type u allowed in the model.

Then the number of states per stage is equal to $\prod_u (x^{\max,u} + 1)$. In total, this sums up to $(T - 1) \cdot \prod_u (x^{\max,u} + 1) + 1$ states,[1] with T denoting the optimization horizon in years. Allowing ten new coal units and ten new CCGT units while considering an optimization horizon of 20 years, this simple formulation leads to 121 states per stage and 2,300 states in total.

---

[1] At the first year, there is only one state, therefore we have to use $T - 1$ instead of T.

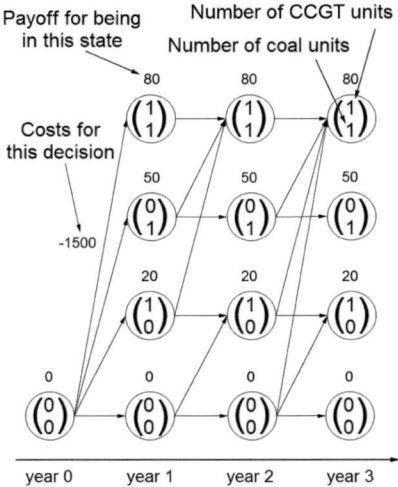

Figure 3.1: A simple dynamic programming representation considering a construction lead time of one year.

Deterministic State Representation Considering Construction Lead Time

One of the unrealistic assumptions in the before mentioned simple example is the non-consideration of construction and planning times. Typical construction times of new power plants are two years for new CCGT plants and three years for new coal plants (Konstantin [Kon08, p. 237]). OECD and IEA [OI05] report construction times between two to seven years for CCGT units, and between three to six years for coal units.

To consider construction times, additional attributes must be added to a state for units that are currently under construction. For each year of construction time beyond the first one, an additional attribute for each unit type is required. Changing the construction time of the problem shown in Figure 3.1 from one year to two years, we get the problem depicted in Figure 3.2.

The consideration of construction time significantly increases the state space. Assuming a planning and construction time of four years for coal units and three years for CCGT units, the total number of states for the before mentioned ten new coal units and ten new CCGT units is 286,286 per stage.[2]

---

[2]There is no simple formula which can be used to determine the number of states per stage. The formula $\prod_u \left((x^{\max, u} + 1)^{t_u^{\text{constr}}} \right)$ with t_u^{constr} being the construction time of unit u forms only a very loose upper bound for the number of states. This is because there are additional constraints for the possible values of the attributes. If we allow a maximum number of ten units per type, and ten

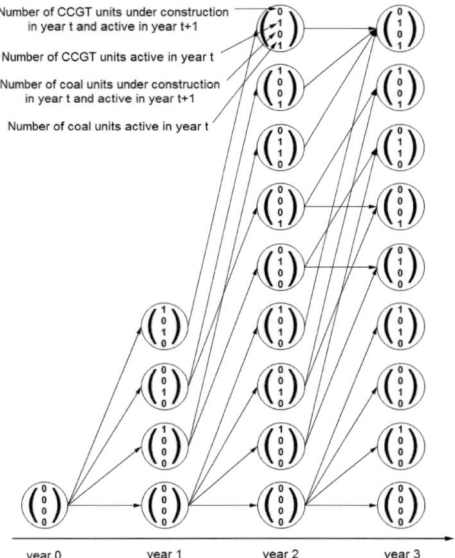

Figure 3.2: A simple dynamic programming representation considering a construction lead time of two years.

Depending on the number of allowed investments, the consideration of construction times of units makes the problem computationally intractable. There are three different possibilities to cope with this problem:

- The simplest one is to ignore construction times. This approach is followed for example by Klein et al. [Kle+08]. The disadvantage of this approach is that it underestimates the effects of uncertainty. When an investment decision is taken, the decision maker in the model has more knowledge than in reality. For example, assume that the construction of a new coal unit with a construction time of four years should be finished in 2010. Ignoring construction lead time, the decision to build this unit is based on the fuel prices of 2009 (or 2010, depending on whether units are immediately available or the next year). In reality, the decision must be taken in 2006, when fuel prices for the year 2010 are still highly uncertain.

- An alternative approach is to restrict the number of investment possibilities to a small number such that the state space remains tractable. For example,

coal units have already been built, then the other attributes representing coal units currently under construction must be zero. For the given example, using the upper bound results in 19,487,171 states per stage.

Hundt and Sun [HS09] consider a problem where investments in one coal unit and one CCGT unit are allowed.

- Instead of restricting the number of investment alternatives, one can also restrict the points in time when investment decisions can be taken. If the period between two points in time where investment decisions are allowed is at least as long as the construction time of new units, then no additional attributes are required. This approach is followed by Botterud [Bot03].

In our opinion, the first approach is too simplistic if the focus of the model lies on an appropriate consideration of fuel and carbon price uncertainty. The second approach is well suited for shorter time periods. However, if a longer optimization horizon is considered, the restriction of the amount of new units to a low number is not realistic, because a longer optimization horizon requires the replacement of a greater amount of old units. Hence, we opt for the third approach. As the focus of our thesis lies on a strategic level, we believe that allowing new investments only every four years is the least constraining of the presented alternatives.

To implement this approach, we simply extend the period a state covers in our model. Instead of considering states representing one year, a state in our model represents four years. This approach is called temporal aggregation. We now have to distinguish between years and stages in the SDP problem. Therefore, we introduce the index s, which denotes the stage in the SDP problem.

Stochastic Fuel Prices

The representations presented so far are based on the assumption that all input data is known with certainty. To account for fuel and carbon price uncertainty, additional attributes must be added to a state.

There are different possibilities how stochastic fuel prices can be considered. Klein et al. [Kle+08] define three different scenarios: high prices, normal prices and low prices. In the high price scenario, all stochastic prices (coal, gas, and carbon) are assumed to be high. Such scenarios aggregating all stochastic parameters have the advantage that only one additional state attribute is required.

Though fuel prices are correlated, they are not perfectly correlated. As a consequence, it is more realistic to consider the different prices individually. For each stochastic price, an additional attribute must be added to a state.

In the following, we denote the set of stochastic fuel and carbon prices with $\mathcal{F}$. Elements of $\mathcal{F}$, e.g., *coal*, *gas* or *carbon* are denoted with f. For a better readability, we often refer to stochastic fuel prices in the following, instead of stochastic fuel and carbon prices. The value of the fuel price scenario for fuel f is denoted with x^f.

Consideration of External Investments

In a competitive market, investors behave differently than in a monopoly, where it can be beneficial to limit the amount of investments to gain higher prices. To prevent that the considered investor behaves like a monopolist, investments of competitors have to be considered. There are two different approaches, how investments of competitors can be considered: they can either be assumed to be independent of the state in the SDP problem, or they can depend on the state.

The first approach assumes that the investments of competitors are known with certainty and that the they are independent of the investment decisions of the considered operator. In this case, these units can be added to the existing power plant fleet when they are built. As these external investment decisions are fixed, no new attributes have to be added to a state.

However, this approach is too simplistic. Several simplifying assumptions are made. First of all, it is assumed that the decisions are independent of fuel prices. Furthermore, it is assumed that investments of competitors are independent of the decisions of the considered investor, and hence the total capacity in the market. Both assumptions are unrealistic and can have strange effects. For example, a competitor may build completely uncompetitive CCGT units in case of high gas and low coal prices.

A more realistic alternative is to model investments of competitors depending on the current state of the problem. There are different possibilities how this can be done. We present two possible approaches later on in this chapter in Section 3.3.

To be able to model external investments depending on the current state and/or the current decision, an additional attribute must be added to a state for each type of new unit that can be built by competitors. As for the attributes representing the new units built by the considered operator, these attributes count the number of external new units of each type.

Representation in our Model

Based on the given descriptions, a state in our model consists of three different groups of attributes: the number of own new units $x^{u^{Own}}$, the number of new units built by competitors $x^{u^{Ext}}$, and the fuel prices x^f modeled as stochastic parameters. Note that $x^{u^{Own}}$, $x^{u^{Ext}}$, and x^f are vectors.

Assuming that we consider investments in CCGT units and coal units, and that we model the coal price, the gas price, and the carbon price as a stochastic parameter, then a state x is represented as follows:

$$x = \begin{pmatrix} x^{u^{Own}} \\ x^{u^{Ext}} \\ x^f \end{pmatrix} = \begin{pmatrix} x^{\text{CCGT unit}^{Own}} \\ x^{\text{Coal unit}^{Own}} \\ x^{\text{CCGT unit}^{Ext}} \\ x^{\text{Coal unit}^{Ext}} \\ x^{\text{Coal}} \\ x^{\text{Gas}} \\ x^{CO_2} \end{pmatrix}.$$

Profits

In Section 1.1.1, we have presented four different markets, on which electricity generation companies can earn profits: the electricity spot market, the electricity derivatives market, the reserve market, and the capacity market. In principle, any of these markets can be considered in our investment model.

However, in this thesis we consider only revenues for the production of electricity. As the electricity prices modeled by our fundamental electricity market model correspond to spot prices, we only consider the spot market. This choice is motivated as follows: for the investment alternatives we consider, coal units and CCGT units, revenues obtained for the production of electricity are by far the most important revenues. Revenues for the supply of control power are more important for peak load units. As the fundamental electricity market models we use do not allow an appropriate determination of control power prices, we disregard reserve markets.[3] Capacity markets do not exist in Germany, hence we do not take them into account in our case studies. Nevertheless, the model can be easily extended to consider capacity remuneration, as it is done by Botterud [Bot03].

---

[3] One possibility to derive control power prices in LP based fundamental electricity market models would be to take the dual values of the control power restrictions (2.3.1). However, prices determined this way deviate significantly from observed prices.

The total reward $R(x)$ of a state corresponds to the discounted sum of rewards obtained for electricity production $R_t^{Prod}(x)$ during each year belonging to the state. These rewards are calculated by one of our fundamental electricity market models. In addition, we must consider the fixed costs $c^{Fixed}(x)$ of the units. As fixed costs are irrelevant for the unit commitment, they are not considered by our fundamental electricity market models. However, they must be taken into account for the investment decisions. Hence, we have to subtract the fixed costs from the annual rewards.

As a state consists of several years, all values have to be discounted to the first year of the state, t_s^1. This is done by applying the discount factor β. The total expected reward for state x is then:

$$R(x) = \sum_{t=t_s^1}^{t_s^1+T^s-1} \beta^{t-t_s^1} \cdot \left(R_t^{Prod}(x) - c^{Fixed}(x) \right),$$
(3.1)

with T^s denoting the number of years per stage.

3.1.2   Decisions

The representation of decisions in our SDP model is straightforward. In this subsection, we briefly outline how decisions are represented in our model and how the costs associated to a decision are determined.

Representation

The decision of an investor is the number of new units of each considered type which he wants to build. To keep the problem computationally tractable, we define an upper bound for the number of new units of each type that can be built. Such a limitation is—if the upper bound is not chosen too low—realistic. It can be interpreted as a regulatory requirement to prevent market power.

As a consequence of the limitation of the maximum number of new units, decisions are state dependent. For example, being already in a state with the maximum number of allowed new units, only the decision to build no new units is possible. We denote the set of possible decisions depending on the state with $\Gamma(x)$, while we use a to refer to decisions.[4]

---

[4]In dynamic programming, either d or a is commonly used for decisions. The abbreviation a stands for *action*.

Thereby, a is a vector of dimension equal to the number of different types of new units that can be built. If investments in coal units and CCGT units are considered, then a decision has the following representation:

$$a = (a^{u^{\text{Own}}}) = \begin{pmatrix} a^{\text{CCGT unit}^{\text{Own}}} \\ a^{\text{Coal unit}^{\text{Own}}} \end{pmatrix}.$$

Costs

Associated with each decision are the costs for building new units. The investment costs per unit might depend on the stage. For example, we might use lower investment costs in later years to represent technological progress. We assume that the investment costs are known with certainty.

As the time horizon of the optimization problem can be shorter than the useful life of the new units, the investment costs for the new units are only considered proportionally. The share of investment costs associated to years considered in the investment model depends on the chosen depreciation method. We define these proportional investment costs $c^{\text{Invest}}(a)$ of a decision at stage s as:

$$c^{\text{Invest}}(a_s) = \sum_{u^{\text{Own}}} \left(a^{u^{\text{Own}}} \cdot \sum_{t=t^1_{s+1}}^{T} ADV(u^{\text{Own}}, t) \right),$$

with $ADV(u^{\text{Own}}, t)$ denoting the annual depreciation value (per unit) depending on the type of new unit and the year. As we assume that new units are operable at the first year of the next stage, we just sum up the annual depreciation values for each unit from the first year of the next stage to the end of the optimization horizon. Then, we multiply this sum by the number of new units of the corresponding type that are built.

To determine the annual depreciation values, different depreciation methods can be used. In the following, we briefly outline two popular approaches, the straight-line depreciation and the sum-of-years' digits method.

- **Straight-line depreciation:** The simplest depreciation method is straight-line depreciation. The straight-line depreciation distributes the investment costs evenly over the expected useful life of a plant. The annual depreciation value is calculated as

$$\text{annual depreciation value} = \frac{\text{investment costs - scrap value}}{\text{life span (years)}}.$$

Straight-line depreciation is appropriate if generated revenues are approximately evenly distributed over the life span. For power plants, this assumption is questionable.

During the first years of their operation, new plants are in general more efficient than most other plants. As time goes by, additional new plants are built. Due to technological progress, these newer plants are more efficient than the plants built before. At the same time, older, less efficient plants may be shut-down. The position of the new plant in the merit–order curve shifts from the left (where the relative cheap units are located) to the right. As a consequence, the amount of generated electricity may decrease with time. This reduced generation tends to result in a lower contribution margin.

- **Sum-of-years' digits depreciation:** To consider these effects, a depreciation method based on decreasing returns can be used. These methods associate higher costs to early periods. A example for such a method is the sum-of-years' digits method.

  To determine the annual depreciation value, the sum of the years' digits is calculated. For a useful life of 5 years, the sum of digits is 5 + 4 + 3 + 2 + 1 = 15. The relative depreciation value of each year is calculated dividing the number of remaining years of useful live (including the current year) by the calculated sum of digits. For the given 5-year example, the relative depreciation value for the first year is 33.3%, for the second year 26.7%, for the following years 20.0%, 13.3% and 6.7%.

Figure 3.3 shows relative accumulated depreciation values for straight-line depreciation and the sum-of-years' digits depreciation. Especially for decisions at the end of the planning horizon (when only few years of the new plants operating time are considered within the optimization), the choice of the depreciation method has a great impact. If only the first five years of operation fall into the planning horizon of our investment problem, the linear depreciation method assigns 16.7% of the investment costs to these five years, while the sum-of-years' digits method considers 30.1% of the costs.

The choice of the depreciation method can be used to reflect the risk-aversion of an investor. A risk-averse investor would choose a depreciation method with relatively high depreciation values for the first years of operation. A risk-neutral investor might use a depreciation method with more evenly distributed depreciation values.

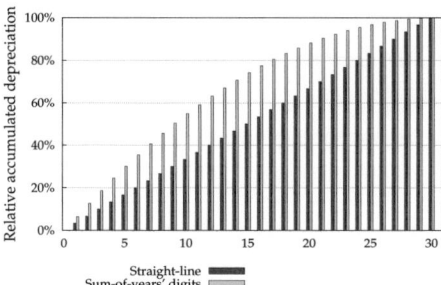

Figure 3.3: Relative accumulated depreciation values depending on the number of years of operation for the straight-line depreciation and the sum-of-years' digits method.

3.1.3 Transitions

The transitions from one state to another in our SDP problem depend on three parts: the decision of the considered operator to build new units, the decisions of competitors to build new units, and the simulated fuel prices. In the following, we describe the effect of these parts on the transitions to a new state.

- **New units built by the considered operator:** As the number of years belonging to one state is chosen such that the construction time of new units is equal to or less than this number, new units are immediately available at the next stage. This part of the transition is deterministic, it only depends on the decision of the considered operator. For the attributes of a state representing the number of units built by the considered operator, the transition function is the following:

$$x_{s+1}^{u^{\text{Own}}} = x_s^{u^{\text{Own}}} + a_s^{u^{\text{Own}}}.$$

- **New units built by competitors:** The number of new units built by competitors is determined by an investment module described in the Section 3.3. Let us denote the external investment decisions at stage s depending on the current state x_s and the current decision a_s with $a_s^{\text{Ext}}(x_s, a_s)$. Then, the transition function for the number of external units is

$$x_{s+1}^{u^{\text{Ext}}} = x_s^{u^{\text{Ext}}} + a_s^{u^{\text{Ext}}}(x_s, a_s).$$

Note that we assume that the investments by competitors depend on the current state as well as the current decision of the considered operator.

- **Fuel prices:** The transition from one fuel price scenario to another is stochastic in our model. We denote the transition probability from fuel price scenario x_s^f to scenario x_{s+1}^f $P(x_{s+1}^f{}' \mid x_s^f)$. In the next section, we describe how we derive these transition probabilities.

3.2   Representation of Fuel Price Uncertainty

In the previous section, we gave a general description of the SDP representation that we use. While we described how stochastic fuel and carbon prices are considered, we did not explicitly specify, how we model these prices. In this section, we describe how we use a Monte Carlo simulation of fuel and carbon prices to derive different scenarios for our SDP model.

There are basically two possibilities how fuel and carbon prices can be modeled. Either one defines a set of scenarios, e.g., low, medium, high, based on assumptions of future prices, or one simulates fuel prices as a stochastic process, and selects some representative scenarios. The first approach is followed for example by Jensen and Meibom [JM08] and Klein et al. [Kle+08], the latter one is used by Hundt and Sun [HS09]. After the simulation of fuel prices, Hundt and Sun use a scenario reduction method to select some representative fuel prices.

While we also use a Monte Carlo simulation of fuel prices, our approach is slightly different from the one followed in [HS09]. For every state, we define an interval for each stochastic fuel price that should be represented by the state. These intervals are based on the fuel prices of the first year of each stage and correspond to the fuel price scenarios x^f used for the states in our SDP model.[5] While we use intervals of 1 €/MWh for fuel prices and 1 €/t for carbon prices, the exact size and the number of intervals might also be chosen differently. For our representation, an example for a state representation is:

$$x^f = \begin{pmatrix} x^{\text{Coal}} \\ x^{\text{Gas}} \\ x^{\text{CO}_2} \end{pmatrix} = \begin{pmatrix} [8.50; 9.50) \\ [16.50; 17.50) \\ [21.50; 22.50) \end{pmatrix}.$$

To determine the probabilities of a state, we use a Monte Carlo simulation to generate a great number, e.g., 50,000, of fuel price paths for the whole optimization

---

[5]Note that we have to define states by the values of the first year of the stage, as the information of the following years is not yet available (e.g., the fuel price development). Therefore, we cannot use for example average fuel prices to characterize a state.

horizon. We can use any stochastic process that is Markovian for the simulation. In our tests we either use a GBM or a trended mean-reverting process.

The simulated fuel price scenarios are then mapped to the corresponding states. For example, the simulated fuel prices $(8.52, 17.24, 22.50)^T$ would be mapped to the previously given state, as well as the scenario $(9.43, 16.84, 21.57)^T$. Based on the mapping of the simulated fuel price scenarios to the states, we can determine the transition probabilities between the states in our SDP model.

Let $\mathcal{I}(x_s^f)$ be the set of the indices of all scenarios that are mapped to x_s^f, where s denotes the stage of the considered state. Then, we can determine the transition probability from x_s^f to $x_{s+1}^f{}'$ as the number of scenarios that belong to both x_s^f and $x_{s+1}^f{}'$ divided by the total number of scenarios belonging to x_s^f:

$$P(x_{s+1}^f{}' \mid x_s^f) = \frac{|\mathcal{I}(x_s^f) \cap \mathcal{I}(x_{s+1}^f{}')|}{|\mathcal{I}(x_s^f)|}.$$

Based on this representation of fuel price uncertainty, a state is not characterized by one fuel price scenario as it is usually the case, but it represents possibly many different scenarios. Hence, the question arises how the contribution margin for a state is calculated. A consideration of all fuel price scenarios is computationally intractable, while a determination of contribution margins based on average fuel prices does not necessarily yield accurate results. This problem is mainly caused by the introduction of temporal aggregation. While for the first year of a state, fuel prices are within the relatively small interval, fuel prices evolve stochastically during the following years of a state. Especially during the last years of a state, the fuel price intervals get quite large and make a determination of contribution margins based on average fuel prices inappropriate.

Another issue of the chosen representation is that the choice of small fuel price intervals results in many different fuel price scenarios, and hence significantly increases the state space of our problem.

In the following parts of this section, we address these two issues. We first propose a method to reduce the number of fuel price scenarios considered in the SDP problem. For this purpose, we aggregate the three different stochastic prices, namely the gas, the coal and the carbon price, into one single measure based on the cost difference between CCGT units and coal units. Next, we illustrate the difficulties related to the determination of the contribution margin of a state with a simple example. To account for these difficulties, we select for every state several representative fuel price scenarios that are used to determine the expected contribution margin of a state.

3.2.1   Aggregation of Fuel Price Scenarios

Despite the temporal aggregation, the state space of the problem might still be too huge, depending on the chosen parameters like the number of new units or the optimization horizon. If an LP-based fundamental electricity market model is used to determine the contribution margins, then even relatively small state spaces can cause very long running times of the model. For that reason, we propose a second form of aggregation based on aggregation fuel prices. Instead of considering stochastic coal, gas and carbon prices explicitly for the state representation, we might use the resulting cost difference between coal units and CCGT units as a state attribute. We refer to this cost difference as *climate spread*.

Let us briefly motivate this choice, before we define how we calculate the climate spread. With the investment model described in this thesis, we focus on a comparison between new coal units and new CCGT units. The unit commitment determined by our fundamental electricity market model depends mainly on the variable costs of a unit compared to other units, not on the absolute costs of a unit. The (conventional) units with the lowest variable costs are nuclear plants, followed by lignite units. Based on current fuel prices, coal units come next, while gas-fired units have the highest variable costs (besides the few oil units). However, depending on the future fuel price development, especially the rank order between coal units and CCGT units might change. Such a change—and the resulting differences in the electricity production—can be captured by the climate spread.

Before we define the climate spread, two other indicators must be introduced, the *clean dark spread (CDS)* and the *clean spark spread (CSS)* (see Caisse des Dépôts [Cai07]). The CDS is the revenue generated by selling power produced by coal units subtracting fuel and carbon costs. Formally defined, the CDS is

$$CDS = p^{\text{Elec}} - (c^{\text{Coal}}/\eta^{\text{Coal}} + c^{\text{CO}_2} \cdot FE^{\text{Coal}}).$$

Thereby, p^{Elec} denotes the electricity price, c^{Coal} the coal price, c^{CO_2} the carbon price, η^{Coal} the efficiency of a coal unit and FE^{Coal} the CO_2 emission factor of a coal unit (in t CO_2/MWh).

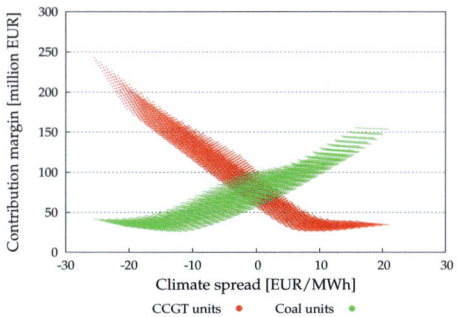

Figure 3.4: Contribution margins of a new CCGT unit and a new coal unit in 2008 depending on different fuel prices and carbon prices.

The CSS is the analogous indicator for CCGT units, defined as

$$CSS = p^{\text{Elec}} - (c^{\text{Gas}}/\eta^{\text{Gas}} + c^{\text{CO}_2} \cdot FE^{\text{Gas}}).$$

We can now define the climate spread CS as difference between the CDS and the CSS. In addition, we account for operation and maintenance costs, which are assumed to be variable costs depending on the electricity production in our model. Then, the climate spread is

$$CS = (CDS - c^{\text{O\&M,Coal}}) - (CSS - c^{\text{O\&M,Gas}}).$$

Following this definition, the climate spread corresponds to the difference in variable costs between coal units and CCGT units. If its value is positive, electricity generated by coal units is cheaper. If the value is negative, CCGT units are more competitive. Coal units will produce more electricity in case of a positive climate spread. Consequently, revenue generated under positive climate spread tends to be higher for coal units than under negative climate spread.

Note that the CDS as well as the CSS depend on the assumed efficiencies η^{Coal} and η^{Gas} for coal and CCGT units. In [Cai07], an efficiency of 40% for coal units and 55% for CCGT units is used. We do not adopt these values, but use the efficiencies of new units (the chosen values are described in Section 5.1).

Figure 3.4 shows the contribution margin of a new coal unit and a new CCGT unit for different fuel and carbon prices in 2008. To derive these values, the LDC model with piecewise-linear adjustments (described in Section 2.1) was used. We took the average fuel prices of 2008, and varied the coal price, the gas price, and the carbon price within an interval of ±50% of their original value. Increments of 5% of the respective prices have been made independently of the other prices.

Two things should be retained from Figure 3.4. First, contribution margins are highly correlated with climate spreads. Average correlation values for 2008 and 2009 are -0.94 for CCGT units and 0.93 for coal units. These values are significantly higher than the correlation with fuel prices or carbon prices.

However, despite the high correlation, representing the contribution margin of a climate spread value with a single value is clearly a simplification. The interval in which a contribution margin lies for a fixed climate spread is relatively wide. To account for this problem, we use several representative fuel price scenarios to determine the contribution margin of a state represented by the climate spread. This approach is described later on in this section.

3.2.2   Difficulties Caused by Temporal Aggregation

While temporal aggregation helps us to significantly reduce the state space, it causes some problems concerning the determination of contribution margins. The reason is that the fuel price scenarios represented by a state can develop differently during the years belonging to the state. If fuel prices vary only within a small interval, resulting deviations in contribution margins are relatively modest. Consequently, the expected contribution margin for all fuel prices within such a small interval might be approximated by the contribution margin obtained with average fuel prices. If fuel price intervals get larger, contribution margins may vary significantly. Due to the non-linear relationship between fuel prices and contribution margins, the contribution margin of average fuel prices does in general not correspond to the average contribution margin considering all fuel price scenarios of a state.

Figure 3.5 illustrates this problem. Figure 3.5(a) shows the results of 1,000 coal price simulations (all other parameters have been fixed). In the first year, coal prices are within the interval [8.5 €/MWh, 9.5 €/MWh). In the forth year, coal prices vary between 3.42 €/MWh and 18.45 €/MWh.

Figure 3.5(b) depicts the resulting contribution margins for one coal unit and one CCGT unit (the contribution margins of both units are summed up). During the first year, the contribution margins are between 82 and 91 million €. In the forth year, the interval grows to 65–199 million €.

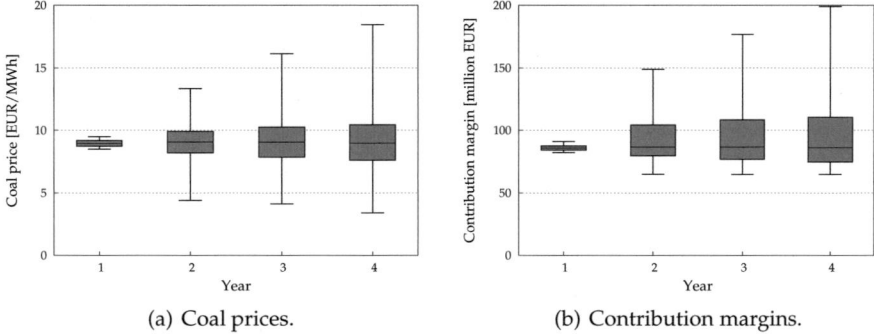

| (a) Coal prices. | (b) Contribution margins. |

Figure 3.5: Boxplot chart for 1,000 coal price simulations and corresponding contribution margins for one coal and one CCGT unit. Bottom and top of the box are the lower and upper quartiles, the ends of the whiskers correspond to minimum and maximum values.

The variance and the skewness of both—coal price and contribution margin—distributions increase with time. In the first year, the contribution margin of average coal prices as well as the average contribution margin (of all coal price scenarios) are 86 million €. In the forth year, the contribution margin of average coal prices is 88 million €, while the average contribution margin of all coal price scenarios is 95 million €. These deviations get even greater, if we do not fix other fuel prices as it was done for the ease of simplicity in this example.

Due to this problem, the determination of the expected contribution margin for a state based on average fuel prices does not deliver accurate results. For this reason, we select several representative fuel prices to calculate the expected contribution margin. The way we determine these representative fuel prices is described in the following subsection.

3.2.3   Choice of Representative Fuel Prices

The computation of the contribution margins for all simulated fuel price scenarios is computational intractable. Consequently, we have to select some representative scenarios for each year of a stage. For this purpose, we use the *fast forward selection*, a popular method in stochastic programming to select representative scenarios. Before we can apply the fast forward selection, we have to answer two questions:

- Should the simulated prices for the different fuels (coal, gas) and carbon be treated together or independently?

Simulation run	Gas price [€/MWh]	Coal price [€/MWh]	Carbon price [€/t]
1	20	8	20
2	16	7	21
3	17	6	16

Table 3.1: Example of fuel price simulations used to illustrate different possibilities to select representative fuel prices.

- Should the simulated fuel and carbon prices be treated as complete time series, or can we consider each year independently?

In the following, we first discuss these two questions. Then, we define a distance metric, which we use for the selection of representative scenarios. Next, we present the algorithm that we use to select these scenarios based on an explicit representation of fuel prices. Finally, we briefly outline how the algorithm is adjusted if the climate spread is used instead.

3.2.3.1   Treatment of Different Fuels

The question we face here is whether we can reduce the simulated prices for the different fuels and the carbon price independently of each other. Let us first illustrate the problem with a simple example.

Example. *Assume that there are three different fuel price simulations as shown in Table 3.1. Selecting a representative scenario for each fuel (and carbon) separately, we would select for each fuel the scenario that lies between the two other scenarios. The result is a gas price of 17 €/MWh, a coal price of 7 €/MWh, and a carbon price of 20 €/t.*
If we consider the coal price, the gas price and the carbon price together, this solution is not feasible as the prices do not belong to the same simulation run. To be able to select representative scenarios, we need a distance metric to determine the distances between the fuel price scenarios of the different simulation runs. Assume that we use the Manhattan distance as distance metric. Then, we would choose the second simulation run. Note that this choice results in a worse approximation of the gas price and the carbon price in this example.

Considering different types of fuels and carbon together reduces the degrees of freedom for the scenario selection. The set of scenarios that can be chosen is a subset of the set of scenarios that can be chosen considering each fuel and carbon separately. Consequently, the solution of a combined consideration cannot be a better representation of the fuel prices than a separate consideration.

Nevertheless, we argue that different fuels and carbon must be considered together for the selection of representative scenarios. There are two reasons for this. First, our true objective is to find a good approximation of the expected contribution margin, not of the fuel prices. The contribution margin of a state is obtained by optimizing the power market operation. This power market operation depends on the value of all fuel prices as well as the carbon price.

Secondly, fuel prices are correlated.[6] If gas prices are high, coal prices tend to be also high. If we select representative fuel prices independently, we lose this relationship between the different fuel prices. Therefore, we consider complete fuel price scenarios consisting of prices of all fuels and the carbon price for the selection of representative scenarios.

3.2.3.2 Treatment of Different Years

The second decision to take is whether complete time-series should be treated together. If complete time-series are treated together, the selected scenarios of the different years of a state must belong to the same simulation run, e.g., if fuel price scenario 4 is selected for the second year of the stage, it must also be selected for the other years. If we do not consider complete time-series, we may choose for example fuel price scenario 4 for the second year, and scenario 8 for the third year. Considering complete time-series may be appropriate when current fuel prices have an important impact on future fuel prices. This is especially the case, if there is a lagged correlation in prices (as it can be observed between gas and oil prices for example). However, we do not consider such lagged correlations.

While the fuel price for the next year depends on the current price,[7] we do not think that this argument is strong enough to consider complete time-series together. Power market operation in year $t + 1$ is independent of the power market operation in year t in our model. As a consequence, fuel prices in year t have no impact on the power market operation in year $t + 1$. Hence, the problem of approximating the contribution margins of a complete state can be divided into approximating the contribution margins of each year separately. Treating each

---

[6]See, e.g., Weber [Web05, p. 58]

[7]This holds for GBMs and mean-reverting processes, two stochastic processes commonly used to simulate fuel prices. Other stochastic processes like ARMA models for example, may use a longer price history. As we use either a (correlated) GBM or a (correlated) mean-reverting process, we do not consider processes relying on a longer price history.

year independently of the others allows for a better approximation, as the decision space for considering all time-series together is a proper subset of the decision space for treating each year separately.

3.2.3.3 Distance Metric

As we consider the different fuel prices of a year together for the scenario selection, a distance metric for fuel price scenarios has to be defined. A simple possibility would be to use the Manhattan distance or the Euclidean distance. However, the application of one of these distance metrics without any adjustments has an important drawback in our opinion.

As already pointed out, our objective is to find a good approximation of the contribution margins for a state. The impact of the different fuel prices on the contribution margin depends on many factors and cannot be exactly predicted. However, the influence of the coal price, the gas price and the carbon price is not the same. This is due to the different units of fuel and carbon prices as well as the different efficiency of CCGT units and coal units.

Consider the case of coal units: An average coal unit with an efficiency of 40% requires 2.5 MWh of coal to produce 1 MWh electricity. Thereby, it emits about $0.86\ t$ of CO_2. The fuel price has a stronger impact on costs compared to the carbon price. Using a distance metric without adjustments does not consider the different impact these prices have on the costs of a unit, and with it on the contribution margin.

We propose to consider these differences in the definition of the fuel price distance function. However, as the efficiencies and the emissions factor of units differ, the choice of the adjustment weights is not obvious. The values we use are based on average efficiencies taken from Caisse des Dépôts [Cai07]. The average efficiency of coal units is assumed to be 0.4, those of CCGT units 0.55. To adjust carbon prices, we take the average of the emission factor for CCGT units ($0.36\ t\ CO_2/MWh$) and the one of coal units ($0.86\ t\ CO_2/MWh$). With these values, we define the distance $D(c_a^{\text{Fuel}}, c_b^{\text{Fuel}})$ between fuel price scenarios c_a^{Fuel} and c_b^{Fuel} as:

$$D(c_a^{\text{Fuel}}, c_b^{\text{Fuel}}) :=$$
$$\sqrt{\left(\frac{(c_a^{\text{Coal}} - c_b^{\text{Coal}})}{0.4}\right)^2 + \left(\frac{(c_a^{\text{Gas}} - c_b^{\text{Gas}})}{0.55}\right)^2 + \left((c_a^{CO_2} - c_b^{CO_2}) \cdot 0.61\right)^2}. \quad (3.2)$$

3.2.3.4 Algorithm

We use Algorithm 1 (which is based on the fast forward selection as described e.g., by Gröwe-Kuska et al. [GK+03]) to determine the representative fuel price scenarios for each state. Note that if two states x_a and x_b differ only in the number of new units x^u, but not in the fuel price scenarios x^f, then we do run Algorithm 1 only once.

Let us briefly comment the algorithm. To select representative fuel prices for a state, we consider all fuel price scenarios associated to this state, $C^{\text{Fuel}}(x)$. For each year of the state, the algorithm iteratively selects a fuel price scenario, until the predefined number of selected scenarios per year, R, is reached.

We start by setting the set of non-selected scenarios $C_t^{\text{Fuel,NonRep}}(x)$ for the current year equal to the set of all fuel price scenarios for this year (line 2). The set of selected fuel price scenarios $C_t^{\text{Fuel,Rep}}(x)$ is empty at the beginning of the algorithm (line 3). At each iteration, a fuel price scenario c_s^{Fuel} is selected such that it minimizes the weighted sum of the distance between all non-selected scenarios to one of the already selected scenarios or to c_s^{Fuel} (lines 10, 21).

We define $d(c_j^{\text{Fuel}}, c_i^{\text{Fuel}})$ as the minimum distance between scenario c_j^{Fuel} and a scenario of the set of selected scenarios or scenario c_i^{Fuel}. During the first iteration, $d(c_j^{\text{Fuel}}, c_i^{\text{Fuel}})$ corresponds to the distance between c_i^{Fuel} and c_j^{Fuel} (line 6), as the set of selected scenarios is empty.

Then we calculate for each non-selected scenario c_i^{Fuel} the weighted sum of minimum distances from all non-selected scenarios to one of the selected scenarios or c_i^{Fuel} (line 8). Thereby, $P(c_j^{\text{Fuel}})$ denotes the occurrence probability for fuel price scenario c_j^{Fuel}. The sum of weighted distances, $z_{c_i^{\text{Fuel}}}$, corresponds the sum of minimum distances between all non-selected scenarios and all selected scenarios under the assumption that c_i^{Fuel} is added to the selected scenarios.

Next, we select the scenario with the minimum sum of weighted distances (line 10), remove this scenario from the set of non-selected scenarios (line 11) and add it to the set of selected scenarios (line 12).

In the subsequent iterations, we only have to adjust $d(c_j^{\text{Fuel}}, c_i^{\text{Fuel}})$. If the scenario chosen during the last iteration, $c_{s_{r-1}}^{\text{Fuel}}$, is closer to c_j^{Fuel} than any of the previously chosen scenarios and c_i^{Fuel}, then $d(c_j^{\text{Fuel}}, c_i^{\text{Fuel}})$ is updated with the distance between c_j^{Fuel} and $c_{s_{r-1}}^{\text{Fuel}}$. The other steps (line 19 - 23) are the same as the ones in the first iteration.

Algorithm 1 Fast forward selection of representative fuel price scenarios

---

1: **for** $t = t_s^1$ **to** $t_s^1 + T^s$ **do**

2:     $\mathcal{C}_t^{\text{Fuel,NonRep}}(x) := \mathcal{C}_t^{\text{Fuel}}(x)$

3:     $\mathcal{C}_t^{\text{Fuel,Rep}}(x) = \varnothing$

4:     **for all** $c_i^{\text{Fuel}} \in \mathcal{C}_t^{\text{Fuel,NonRep}}(x)$ **do**

5:        **for all** $c_j^{\text{Fuel}} \in \mathcal{C}_t^{\text{Fuel,NonRep}}(x)$ **do**

6:          $d(c_j^{\text{Fuel}}, c_i^{\text{Fuel}})^{[1]} := D(c_j^{\text{Fuel}}, c_i^{\text{Fuel}})$

7:        **end for**

8:        $z_{c_i^{\text{Fuel}}}^{[1]} := \sum\limits_{c_j^{\text{Fuel}} \in \mathcal{C}_t^{\text{Fuel,NonRep}}(x)} P(c_j^{\text{Fuel}}) \cdot d(c_j^{\text{Fuel}}, c_i^{\text{Fuel}})^{[1]}$

9:     **end for**

10:    $c_{s_1}^{\text{Fuel}} = \underset{c_i^{\text{Fuel}} \in \mathcal{C}_t^{\text{Fuel,NonRep}}(x)}{\operatorname{argmin}} z_{c_i^{\text{Fuel}}}^{[1]}$

11:    $\mathcal{C}_t^{\text{Fuel,NonRep}}(x) = \mathcal{C}_t^{\text{Fuel,NonRep}}(x) \setminus \{c_{s_1}^{\text{Fuel}}\}$

12:    $\mathcal{C}_t^{\text{Fuel,Rep}}(x) = \mathcal{C}_t^{\text{Fuel,Rep}}(x) \cup \{c_{s_1}^{\text{Fuel}}\}$

13:

14:    **for** $r = 2$ **to** R **do**

15:      **for all** $c_i^{\text{Fuel}} \in \mathcal{C}_t^{\text{Fuel,NonRep}}(x)$ **do**

16:        **for all** $c_j^{\text{Fuel}} \in \mathcal{C}_t^{\text{Fuel,NonRep}}(x)$ **do**

17:          $d(c_j^{\text{Fuel}}, c_i^{\text{Fuel}})^{[r]} := \min\left(d(c_j^{\text{Fuel}}, c_i^{\text{Fuel}})^{[r-1]}, d(c_j^{\text{Fuel}}, c_{s_{r-1}}^{\text{Fuel}})^{[r-1]}\right)$

18:        **end for**

19:        $z_{c_i^{\text{Fuel}}}^{[r]} := \sum\limits_{c_j^{\text{Fuel}} \in \mathcal{C}_t^{\text{Fuel,NonRep}}(x)} P(c_i^{\text{Fuel}}) \cdot d(c_i^{\text{Fuel}}, c_j^{\text{Fuel}})^{[r]}$

20:      **end for**

21:      $c_{s_r}^{\text{Fuel}} = \underset{c_i^{\text{Fuel}} \in \mathcal{C}_t^{\text{Fuel,NonRep}}(x)}{\operatorname{argmin}} z_{c_i^{\text{Fuel}}}^{[r]}$

22:      $\mathcal{C}_t^{\text{Fuel,NonRep}}(x) = \mathcal{C}_t^{\text{Fuel,NonRep}}(x) \setminus \{c_{s_r}^{\text{Fuel}}\}$

23:      $\mathcal{C}_t^{\text{Fuel,Rep}}(x) = \mathcal{C}_t^{\text{Fuel,Rep}}(x) \cup \{c_{s_r}^{\text{Fuel}}\}$

24:    **end for**

25: **end for**

3.2.3.5 State Representation Based on the Climate Spread

The algorithm just described is based on the assumption of an explicit representation of the stochastic fuel prices. If we use the climate spread to reduce the problem's dimension, the choice of the representative fuel price scenarios deviates a little bit from the approach outlined before. In principle, we also use Algorithm 1 to select representative climate spread scenarios for each year of the stage. But instead of using the distance metric defined in equation (3.2), we take the absolute difference between two climate spreads values (e.g., for climate spread values of 2.3 and 3.7, we use 1.4 as distance, independent of the concrete fuel prices associated to these climate spreads).

While the use of the climate spread helps us to reduce the complexity of the problem by decreasing the dimensionality, it has its drawbacks. As we have seen in Figure 3.4 on page 125, the same climate spread values obtained with different fuel prices can result in different contribution margins. To account for these problems, we do not only use one fuel price scenario per representative climate spread. Instead, we use Algorithm 1 to select several (for example three or five) representative fuel price scenarios for each representative climate spread scenario. Note that in line 2 of the algorithm, we replace the set of all fuel prices associated to state x with the subset of all fuel prices with the selected climate spread associated to state x.

3.2.4 Determination of Contribution Margins

The selected representative fuel price scenarios are used in our fundamental electricity market model to determine the corresponding contribution margins of the new units. To consider also the other, non-selected fuel price scenarios, we interpolate the contribution margins of these scenarios.

Thereby, we distinguish two cases: If we use the climate spread, we linearly interpolate the contribution margin for a non-selected fuel price scenario by its two nearest neighbors. We opt for the linear interpolation due to its low computational burden.

If we do not use the climate spread, the problem is more complicated. Assuming that we have at least two stochastic parameters (e.g., the gas price and the coal price), the identification of the nearest neighbors of a fuel price scenario is more complex than in the one-dimensional case. Under these circumstances, we use a kernel regression to estimate the contribution margin for the non-selected fuel prices.

Kernel regression is a non-parametric technique that estimates the value of a query point based on the known values of points in the neighborhood of the query point. The weight of each known value in the neighborhood of the query point is determined by the Kernel function. Note that Kernel regression requires us to find the nearest neighbors of the query point. Therefore, it is slower than a linear regression, but its results are in general more accurate. An introduction to kernel regression is given, e.g., by Härdle [Här92].

To estimate the contribution margins of the non-selected fuel price scenarios, we use a Nadaraya-Watson kernel regression (Nadaraya [Nad64], Watson [Wat64]):

$$\widetilde{R}^{\text{Prod}}\left(x, c^{\text{Fuel}}, t\right) = \frac{\sum\limits_{c_r^{\text{Fuel}} \in C_t^{\text{Fuel,Rep}}(x)} \left(K\left(\frac{D(c^{\text{Fuel}}, c_r^{\text{Fuel}})}{h}\right) \cdot R^{\text{Prod}}\left(x^u, c_r^{\text{Fuel}}, t\right) \right)}{\sum\limits_{c_r^{\text{Fuel}} \in C_t^{\text{Fuel,Rep}}(x)} K\left(\frac{D(c^{\text{Fuel}}, c_r^{\text{Fuel}})}{h}\right)},$$

with $R^{\text{Prod}}(x^u, c_r^{\text{Fuel}}, t)$ denoting the calculated contribution margin for the representative fuel prices. To determine the weights of the representative fuel prices, we use the Epanechnikov kernel as Kernel function $K(u)$. The Epanechnikov kernel is defined as

$$K(u) = \frac{3}{4} \cdot (1 - u^2) \cdot \mathbb{1}_{[|u|<1]}.$$

The parameter h is the kernel bandwidth. The smaller the parameter h, the less points are used for the estimation, as more points get a weight of 0.

Based on the contribution margins of the representative fuel prices and the interpolated contribution margins, we determine the expected contribution for state x in year t as:

$$R_t^{\text{Prod}}(x) = \sum_{c^{\text{Fuel}} \in C_t^{\text{Fuel,Rep}}(x)} \left(P(c^{\text{Fuel}} \mid x, t) \cdot R^{\text{Prod}}(x^u, c^{\text{Fuel}}, t) \right)$$

$$+ \sum_{c^{\text{Fuel}} \in C_t^{\text{Fuel,NonRep}}(x)} \left(P(c^{\text{Fuel}} \mid x, t) \cdot \widetilde{R}^{\text{Prod}}(x^u, c^{\text{Fuel}}, t) \right).$$

Thereby, $P(c^{\text{Fuel}} \mid x, t)$ denotes the conditional probability for fuel price scenario c^{Fuel} in year t and state x. The contribution margins calculated by the fundamental electricity market model are denoted with $R^{\text{Prod}}(x^u, c^{\text{Fuel}}, t)$, the interpolated ones are $\widetilde{R}^{\text{Prod}}(x^u, c^{\text{Fuel}}, t)$.

3.3 Investments of Competitors

An important, but difficult task in an SDP based investment model is the inclusion of competitors. Most other SDP based models either ignore the problem of competitors or circumvent it by taking a centralized point of view, optimizing the power generation expansion for the whole market.

Botterud [Bot03] introduces an electricity price cap in his model which he justifies as the effect of new investments by competitors. This approach has the advantage that it does not introduce additional complexity by increasing the state space. However, it has two drawbacks. First, the price cap constitutes only a rough approximation of the effect of new units. Secondly, it has only an effect when prices are high. If—for example due to decreasing fuel prices—electricity prices drop again, the price cap does no longer influence electricity prices. In reality, if new units are built, their influence on the market persists.

In this section, we propose two different approaches how investments of competitors can be considered in our model. The first approach is simply based on the difference between the electricity supply and the maximum annual load. The second approach considers the expected profitability of new plants.

For both approaches, we assume that the investment decisions by competitors are taken after the considered operator has made his decision. New units built by competitors are operable at the beginning of the next stage, e.g., if investments are triggered during stage one, the new units are added to the power generation fleet at the first year of stage two. For the sake of simplicity, we assume in the following that new coal units and CCGT units have the same capacity. The model can be easily adopted to different capacities.

3.3.1 Investments Based on Shortage of Supply

When electricity demand exceeds electricity supply, electricity prices reach extreme peaks. During these times, the normal, marginal cost based mechanism to determine electricity prices ceases to be valid. Under these circumstances, the electricity price depends on the willingness to pay of the demand side, which corresponds to the VoLL. In our electricity market models, we set the VoLL to 2,000 €/MWh.

Such price peaks are considered as necessary investment incentives in literature. In a competitive market, investments are stimulated by such extreme prices. In a monopoly though, the opposite may be the case. The monopolist may try to limit the supply to obtain these extreme high prices. If we do not consider investments of competitors, we can observe such a behavior in our model. The considered operator will build just a few plants to profit from these high prices, but not enough to satisfy the demand.

To avoid such a scenario, we may model investments by competitors based on the scarcity of supply. Thereby, two decisions must be taken: the number of new units that are built by competitors must be determined, and the type of new unit must be selected.

3.3.1.1 Determination of the Number of New Units

In the approach based on the scarcity of supply, we decide on new investments such that for every year the available capacity is greater or equal to the maximum demand plus a predefined margin. Formally written, we determine the total number of new units n_s^{Total} as

$$
n_s^{\text{Total}}(x_s, a_s) = \max\left(0; \left\lceil \max_{t \in \mathcal{T}_{s+1}} \left(\frac{(\text{peak load}_t + \text{margin}) - \text{supply}_t(x_s, a_s)}{\text{capacity per unit}} \right) \right\rceil \right).
\tag{3.3}
$$

Thereby, we denote the set of years belonging to stage $s + 1$ with $\mathcal{T}_{s+1}$. Note that the number of new units is based on the maximum difference between the peak load plus the margin and the supply occurring during one of the years belonging to the next stage.

The margin can be used to simulate different degrees of competition in the market. Using a low value (or even negative) for this margin can be interpreted as a market with a low degree of competition. For a high value, the inverse is the case.

It is important to note that investments based on equation (3.3) are different from a predefined, fixed investment schedule. As we assume that the considered operator takes its decision first, investments by competitors are only carried out, if the considered operator does not invest sufficiently. This is ensured by equation (3.3) as the capacity of new units built by the considered operator is added to the supply in the following years.

3.3.1.2 Determination of the Type of New Units

Besides the amount of new units, we have to decide on what type of new units should be built. One possibility would be to build the same number of CCGT units and coal units. However, such a behavior is not be very realistic. In case of a positive climate spread, one might prefer coal units, while for a negative climate spread, CCGT units are more attractive.

To take these considerations into account, we use the observed CDSs and CSSs to select how many units of each type are built. Remember that the CDS and the CSS correspond to the difference between the electricity price and the variable costs for coal units and the electricity price and the variable costs of CCGT units, respectively.

In our SDP problem, there are different paths—corresponding amongst others to different fuel price scenarios—through which a state can be reached. As the observed CDSs and CSSs depend on the realized path, we cannot use the values of previous stages.

We might use the average CDS and CSS of the current state to decide what kind of units are built. However, as we assume that new units are ready for operation at the first year of the next stage, the construction of the new units must be started before all CDSs and CSSs of the current state are observed. Therefore, we use only the CDS and CSS of the first year of the current state to determine the number of new units of each type.

A drawback of the CDS and CSS is that they consider only variable costs. Investment costs or other fixed costs are not considered. To account for these costs, we calculate a threshold value for the CDS and CSS that must be exceeded to cover these costs. Then, we determine the relative profitability of CCGT units and coal units as the ratio between the observed CDS respectively CSS and the calculated threshold values.

To determine these threshold values, we have to make several simplifying assumptions. First, we assume that a new unit generates constant contribution margins during each year of operation. This allows us to calculate the annual contribution margin R'^{Prod} required to cover investment costs c^{Invest} as

$$R'^{\text{Prod}} \overset{!}{=} \frac{c^{\text{Invest}}}{a_{\overline{n}|i}}.$$

Thereby, $a_{\overline{n}|i}$ denotes the annuity factor,[8] which is calculated as

$$a_{\overline{n}|i} = \frac{1}{i} - \frac{1}{i \cdot (1+i)^n},$$

---

[8] An introduction to the concept of annuities including the derivation of the formula for the annuity factor can be found in most finance textbooks, e.g., Brealey et al. [Bre+07].

	Unit	CCGT	Coal
Capacity	MW	1,000	1,000
Investment costs	M€	500	1,000
Useful life	a	25	35
Required annual contribution margin	M€	42.9	77.2
Annual fixed costs	M€	15	20
Full load hours	h/a	5,000	5,500
Threshold value	€/MWh	11.58	17.67

Table 3.2: Determination of threshold values for external investments based on an interest rate of 0.07. Sources: DEWI et al. [DEW+05], Konstantin [Kon08], own calculations.

with i denoting the risk free interest rate, and n the useful life of the new unit.

To be able to determine a threshold value for the CDS and the CSS, respectively, we further need to make assumptions about the number of full load hours, t^{FL}. Based on the required annual contribution margins, the annual fixed costs and the number of full load hours, we can determine the threshold value for the CDS as

$$CDS^{\mathrm{Threshold}} = \frac{\frac{R'^{\mathrm{Prod}}_{\mathrm{coal}} + c^{\mathrm{Fixed}}_{\mathrm{coal}}}{p^{\mathrm{max}}_{\mathrm{coal}}}}{t^{\mathrm{FL}}_{\mathrm{coal}}}, \tag{3.4}$$

where $p^{\mathrm{max}}_{\mathrm{coal}}$ denotes the capacity of a new coal unit. The threshold value for the CSS is calculated correspondingly as

$$CSS^{\mathrm{Threshold}} = \frac{\frac{R'^{\mathrm{Prod}}_{\mathrm{gas}} + c^{\mathrm{Fixed}}_{\mathrm{gas}}}{p^{\mathrm{max}}_{\mathrm{gas}}}}{t^{\mathrm{FL}}_{\mathrm{gas}}}, \tag{3.5}$$

where $p^{\mathrm{max}}_{\mathrm{gas}}$ corresponds to the capacity of a new CCGT unit. Table 3.2 shows possible threshold values.

To determine the number of new units of each type, we divide the observed annual CSS during the first year of the stage, $CSS_{t^1_s}(x_s)$, and the observed annual CDS during the first year of the stage, $CDS_{t^1_s}(x_s)$, by the corresponding threshold value:

$$\gamma^{\mathrm{CCGT}}(x_s) = \frac{CSS_{t^1_s}(x_s)}{CSS^{\mathrm{Threshold}}},$$

$$\gamma^{\mathrm{coal}}(x_s) = \frac{CDS_{t^1_s}(x_s)}{CDS^{\mathrm{Threshold}}}.$$

Note that for the ease of simplicity, we use the average annual electricity price to determine the spread values. Considering only the hours in which a unit operates to determine the spread values would result in higher spreads (as a unit tends to be shut-down during periods of lower electricity prices). As the determination of spreads based on the operating hours is more complicated, and because the usage of the average annual electricity price has similar effects on both spreads, we use the average annual electricity price to determine the spread values.

The value of γ is an indicator of the profitability of new CCGT units and new coal units, respectively. If γ is greater than one, investments in new CCGT units are profitable under the assumption that the future contribution margins correspond to the one on which the calculation of the threshold value is based. If the value is between 0 and 1, the unit is only able to recover a part of the investment costs. A negative value would signify that a unit is even not able to cover any part of the investment costs.

The assumption of constant contribution margins for the whole useful life of a unit is obviously not realistic. Therefore, we do not only select the more profitable unit type, but use the ratio of the calculated γ to select the number of new units. Formally defined, we determine the number of new CCGT units n_s^{CCGT} as

$$n_s^{\text{CCGT}}(x_s, a_s) = \text{Round}\left(n_s^{\text{Total}}(x_s, a_s) \cdot \frac{\gamma^{\text{CCGT}+}(x_s)}{\gamma^{\text{CCGT}+}(x_s) + \gamma^{\text{coal}+}(x_s)}\right). \tag{3.6}$$

Note that we have to use γ^+, the maximum of γ and 0, to avoid that the number of units of a type gets negative. The number of new coal units n_s^{Coal} is calculated as

$$n_s^{\text{Coal}}(x_s, a_s) = n_s^{\text{Total}}(x_s, a_s) - n_s^{\text{CCGT}}(x_s, a_s). \tag{3.7}$$

The investment decision of competitors is then

$$a_s^{u^{\text{Ext}}}(x_s, a_s) = \begin{pmatrix} n_s^{\text{CCGT}}(x_s, a_s) \\ n_s^{\text{Coal}}(x_s, a_s) \end{pmatrix}.$$

3.3.2 Investments Based on Expected Profitability

The investment rule presented in the previous subsection can be used to avoid that the considered operator exploits the VoLL to generate extreme profits. But still, even if there is enough capacity to satisfy demand, there might be situations in which the considered operator has an incentive to invest much less than it would be the case in a competitive market.

Such a situation can occur for example, if the fuel price development is favorable for one technology. Under such circumstances, electricity produced by the unfavorable technology gets very expensive. In a competitive market, investments in the cheaper technology will take place. In our model with investments by competitors solely based on missing capacity, it might be profitable for the considered operator not to invest in too many units with the cheaper technology, as this reduces electricity prices.

Such a behavior can be avoided, if the investments of competitors are based on the expected profitability of new units. We can use the threshold values for the profitability of new coal units and CCGT units defined in equations (3.4) and (3.5) as a basis for such an investment rule.

If these threshold values are exceeded, investments in new units are profitable. To account for uncertainties or risk aversion, we might require the observed CDS and CSS to exceed the corresponding threshold value by a predefined security margin Δ^{Sec} before investments are triggered. Fore example, investments in CCGT units might be triggered if $CSS_{t_s^1}(x_s) \geq (CSS^{\text{Threshold}} + \Delta^{\text{Sec,CCGT}})$.

Based on this rule, we can decide whether competitors invest or not. However, if investments are profitable, we must somehow decide how many new units are built. Obviously, the greater the difference between observed CDS and CSS, respectively, and the corresponding threshold values, the more profitable are investments in new units. Consequently, a greater difference tends to result in the construction of more new units.

The difficulty is that each new unit has an influence on electricity prices and electricity production. Every new unit reduces the profitability of existing units. The exact effect of a new units depends on multiple factors like current fuel prices and the current power generation fleet. In case of scarcity of supply, the impact of a new unit is the greatest. In such a case, a new unit either reduces the periods in which demand exceeds supply and the VoLL is paid, or it replaces the most expensive units. If electricity supply is much greater than electricity demand, a new unit might only replace a unit that is slightly more expensive than the new unit. Consequently, the average electricity price decreases only marginally. For the sake of simplicity, we assume a constant price effect pe^{Unit} in our model.

Let us denote the amount by which electricity prices are (assumed to be) reduced by a new unit with pe^{Unit}. With this price effect, we determine the total number of new units as

$$n_s^{\text{Total}}(x_s, a_s) = \max\left(0, \text{Round}\left(\frac{ex^{\text{Max}}(x_s)}{pe^{\text{Unit}}} - \sum_{u^{\text{Own}}} a^{u^{\text{Own}}}\right)\right),\qquad (3.8)$$

with the maximum excess ex^{Max} defined as

$$ex^{\text{Max}}(x_s) = \max\Big(CSS_{t_s^1}(x_s) - (CSS^{\text{Threshold}} + \Delta^{\text{Sec,CCGT}}),$$
$$CDS_{t_s^1}(x_s) - (CDS^{\text{Threshold}} + \Delta^{\text{Sec,coal}})\Big).$$

Note that we subtract the number of new units built by the considered operator from the number of new units built by competitors, as we assume that the considered operator takes its decision first. Consequently, the competitors can consider the assumed price effect of these new units. As before, we use the CDS and CSS observed during the first year of the current stage to determine the number of new units.

A disadvantage of this approach is that it leads to lagged external investments. The construction of new units is only started after a high profitability is observed. It seems plausible that at least rising prices caused by scarcity of supply can be anticipated.

Therefore, instead of using only equation (3.8) to determine the number of new units built by competitors, we use the maximum of the number of new units based on the scarcity of supply in equation (3.3) and the number of new units based on expected profitability in equation (3.8). To decide how many new units of each type are built, we still use equations (3.6) and (3.7).

3.4 Interpolation-Based Stochastic Dynamic Programming

In the previous sections we described the representation of our investment model. While we have introduced two kinds of simplifications, temporal aggregation and aggregation of fuel prices, the consideration of external investments increases the problem size. The resulting power generation expansion problem might still be too complex to be solved. Depending on the number of new units which can be built, the optimization horizon, and the number of considered fuel price scenarios, the state space may get too large.

	Base	TA	TA + CS
# permutations of attributes representing units	1.0E+12	14,641	14,641
# permutations of attributes representing prices	125	125	20
Number states per stage	1.26E+14	1,830,125	292,820

Table 3.3: Example for the number of states per stage depending on the representation. Base stands for the model without simplifications, TA denotes temporal aggregation, CS denotes the representation based on the climate spread.

Table 3.3 gives an example for the problem size depending on different state representations. It is based on the assumption that investments in up to ten CCGT units and up to ten coal units are allowed, while the same amount of external investments is considered. The construction time of new units is four years. Concerning the number of fuel price scenarios, we assume that five different scenarios are used for the gas price, the coal price and the carbon price, respectively. As these prices are not perfectly correlated, any combination of these price scenarios may occur, resulting in $5^3 = 125$ different scenarios in case of a fuel price based representation.[9] We furthermore assume that if the climate spread is used, we only need 20 scenarios.

The total number of states per stage is obtained by multiplying the number of considered price scenarios by the number of possible scenarios concerning the number of new units. Especially the temporal aggregation reduces the state space significantly. But even if we apply temporal aggregation and use the climate spread, we still have 292,820 states per stage. The great number of states is undesirable for two reasons: First, we may get memory problems.[10] Secondly, we have to calculate the expected contribution margin for every state. Even if we use the fast LDC model, the determination of the contribution margins might be the bottleneck of the algorithm.

In this section, we present another method to further reduce the complexity. Instead of solving the problem based on the original state space, we solve an approximation of the problem consisting of a compact subset of the state space. While we restrict the number of considered states, we do not restrict the set of possible

---

[9]One may think that we do not need that many fuel and carbon price scenarios. This might be true for the next two or three years. However, if we consider an optimization horizon up to 30 years, 125 scenarios are not that many. For example, 90% of carbon prices in the year 2030 are—based on our simulations—in the interval of 27 €/t–93 €/t. For such an interval, we will probably use more than five scenarios.

[10]Remember that we do not only have to save the states, but also the transition matrix.

decisions. Hence, decisions may lead to non-considered states. For these non-considered states, we interpolate the expected value.

A popular choice for the selection of the considered states is the grid-based approach. Thereby, a grid is placed in the state space. The grid points form the subset of the state space that is used for the optimization. Such grid-based methods are for example commonly used to discretize continuous problems.

Different grid-based methods exist. The most straightforward one is the choice of uniform grids. A more elaborated method are adaptive grid schemes. Instead of equidistant grid points, these methods try to use a finer grid in the interesting region of the state space. The selection of these interesting regions is based on the priority calculated by a splitting criterion. These priorities are used to determine which hyperrectangles of the grid are refined to get a better approximation of the corresponding area of the state space.

Adaptive grid methods allow to significantly reduce the complexity of a problem. Coming back to the example depicted in Table 3.3, we might start with a coarse grid considering three different scenarios for each attribute representing the number of new units, and five scenarios for the climate spread. In this case, we only consider 405 states per stage during the first iteration of the algorithm. The number of states required before the algorithm converges depends on multiple factors. In our tests, we need for problems with the same size as the one described here in most cases only between 1,000–10,000 states per stage before the algorithm converges.

In the following, we start by describing the basic idea of our ISDP approach. Then we briefly introduce uniform grid methods, before we present adaptive grid methods. Next, we compare two methods that can be used to interpolate the value function of non-considered states. These methods are multilinear interpolation and simplex-based interpolation. In the last part of this section, we present and discuss several splitting criteria.

3.4.1 Basic Idea

The basic idea of ISDP is to replace the computationally intractable original problem with a more tractable approximated problem. Instead of calculating the optimal decision a^* and the value function $V(x)$ for all states of the state space $\mathcal{X}$, this is only done for a subset of the state space, $\mathcal{X}^{\text{Saved}}$.

A pseudo-code notation of the backward-induction algorithm applied to our approximated investment problem is given by Algorithm 2. We start by setting the

---

Algorithm 2 Backward induction

1: $V(x_{N+1}) = 0$
2: **for** s = N **down to** 0 **do**
3:     **for all** $x_s \in \mathcal{X}_s^{\text{Saved}}$ **do**

4:
$$V(x_s) = \max_{a_s \in \Gamma(x_s)} \Bigg(F(x_s, a_s)$$
$$+\beta^{T_s} \cdot \sum_{x'_{s+1} \in T(x_s, a_s) \cap \mathcal{X}_{s+1}^{\text{Saved}}} \left(P(x'_{s+1} \mid x_s, a_s) \cdot V(x'_{s+1}) \right)$$
$$+\beta^{T_s} \cdot \sum_{x'_{s+1} \in T(x_s, a_s) \cap \mathcal{X}_{s+1}^{\text{NonSaved}}} \left(P(x'_{s+1} \mid x_s, a_s) \cdot \tilde{V}(x'_{s+1}) \right) \Bigg)$$

5:
$$a_s^* = \operatorname*{argmax}_{a_s \in \Gamma(x_s)} \Bigg(F(x_s, a_s)$$
$$+\beta^{T_s} \cdot \sum_{x'_{s+1} \in T(x_s, a_s) \cap \mathcal{X}_{s+1}^{\text{Saved}}} \left(P(x'_{s+1} \mid x_s, a_s) \cdot V(x'_{s+1}) \right)$$
$$+\beta^{T_s} \cdot \sum_{x'_{s+1} \in T(x_s, a_s) \cap \mathcal{X}_{s+1}^{\text{NonSaved}}} \left(P(x'_{s+1} \mid x_s, a_s) \cdot \tilde{V}(x'_{s+1}) \right) \Bigg)$$

6:     **with:** $F(x_s, a_s) = R(x_s) - c^{\text{Invest}}(a_s)$
7:     **end for**
8: **end for**

---

value function for dummy state x_{N+1} to zero.[11] We assume that x_{N+1} is the (only) successor of all states of the last stage, independent of the chosen decision. Then we iteratively determine the optimal decision and the value function for all selected states of a stage. Thereby, we start with the states of the last stage and move forward stage by stage.

The optimal decision as well as the value function of a state depend on the immediate payoff-function of the current state and decision, $F(x_s, a_s)$, and the expected discounted value of the succeeding states. The payoff function is composed of the investment costs $c^{\text{Invest}}(a_s)$ associated to the investment decision and the (expected) profit $R(x_s)$ for being in x_s (defined in equation (3.1) on page 118).

For the calculation of the expected value of the succeeding states, the (stochastic) transition function $T(x_s, a_s)$ is used. Thereby, $P(x'_{s+1} \mid x_s, a_s)$ denotes the transition probability from state x_s to state x'_{s+1} if action a_s is taken.

If a succeeding state is part of the grid, we have already calculated its value function. If the successor is part of the set of non-selected states $\mathcal{X}^{\text{NonSaved}}$, then we use an approximation of its value function, $\tilde{V}(x'_{s+1})$, which we obtain by interpolation. The expected values of succeeding states are discounted by the discount

---

[11]Salvage values of new units at the end of the optimization horizon are directly deduced from investment costs.

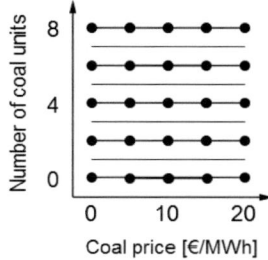

Figure 3.6: Example of a fixed grid considering only coal units.

factor β, which must be taken to the power of the number of years per stage, T^s. While Algorithm 2 outlines the general idea of the approach we use, it does not specify how the set of considered states is selected, and how the expected value of non-selected states is interpolated. We discuss these issues in the following subsections.

3.4.2   Uniform Grid Methods

The selection of the subset of the states that are used for the optimization is often based on grids. The simplest grid based method is the application of a uniform grid. If a uniform grid is used, every saved state has the same distance to its neighbors.

Figure 3.6 depicts a possible uniform grid for a simplified example allowing only investments in coal units. The points represent the subset of the state space used to solve the dynamic programming problem, while the lines represent the complete state space (assuming that coal prices are between 0 and 20 €/MWh and that the maximum number of allowed new units is eight).

The advantage of a uniform grid is its simplicity. The only parameter that has to be chosen is the distance between the saved states. Thereby, a compromise must be found between accuracy of the approximated solution and complexity. On the one hand, a smaller distance results in a better approximation of the value function. On the other hand, it also increases the computational burden.

Especially for higher-dimensional problems, it is difficult to find an acceptable solution using uniform grids. Either the number of grid points is too low to get a good approximation of the value function, or the number of grid points is too high to solve the problem.

Under such circumstances, more intelligent methods are required. The main problem of uniform grids is that they try to approximate the value function over the whole state space with the same accuracy. However, there are regions in the state space that are more important than others. Three different cases in which a coarse grid is sufficient can be distinguished:

- Some regions of the state space are never reached during the optimization. For example, assume that the construction of more than six new coal units always results in losses independent of the fuel prices. Then, it is sufficient to know that the value function is negative in this region. It does not matter whether the losses are 100 million € or 1 billion €.

- While the number of new units is fixed by our decisions, we cannot influence the fuel price development. Hence, we cannot guarantee that some fuel prices are never reached. However, some fuel price scenarios are more probable than others. Therefore, it is more important to have a good approximation for these scenarios compared to unlikely scenarios.

- Finally, we do not need a detailed representation of regions in which the value function can be well interpolated. This is basically the case in regions with a flat value function. For example, if no new units are built, the contribution margin is—independent of the fuel price—zero. Consequently, the value of states on the lowest line in Figure 3.6 can be well approximated. Note however, that it may depend on the interpolation method whether the value function within a region can be well approximated.

3.4.3   Adaptive Grid Methods

Adaptive aggregation methods (e.g., Grüne and Semmler [GS04], Monson et al. [Mon+04], Munos and Moore [MM02]) try to focus on the most important regions of the state space. These methods start with a relatively coarse grid, which is iteratively refined. The refinement of the grid is based on splits of the (hyper)rectangles of which the grid is composed. Figure 3.7 shows an example in 2D.

Algorithm 3 gives a pseudo-code notation of adaptive grid methods. Let us briefly comment the different parts of the algorithm. We start with an initial grid, which may be relatively coarse, as it is iteratively refined. Then, we solve the SDP problem based on the reduced state space as described in Algorithm 2.

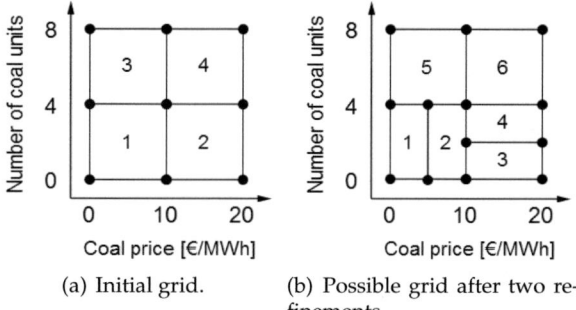

(a) Initial grid.

(b) Possible grid after two refinements.

Figure 3.7: Example of an initial grid and possible refinements.

Algorithm 3 Adaptive grid methods

1: Select initial grid
2: Solve (S)DP problem with initial grid
3: **while** Stopping criterion not fulfilled **do**
4:    Calculate splitting priority and splitting dimension for all (hyper-) rectangles of the grid
5:    Refine the grid
6:    Solve (S)DP problem with refined grid
7: **end while**

After the solution of the problem based on the initial grid, we enter a loop which is iteratively executed until a predefined stopping criterion is met. This stopping criterion may be a fixed number of refinements, or it may be based on the splitting priority for the hyperrectangles of the grid.

If the stopping criterion is not met, we calculate the splitting priority and the splitting dimension for all hyperrectangles of the grid. This is probably the most important part of the algorithm, and is described in detail in Subsection 3.4.5.

Based on the splitting priorities, the grid is refined. We either select a fixed number of rectangles with the highest splitting priorities, or we select all rectangles with a splitting priority exceeding a predefined threshold. If possible, we split the chosen rectangles in the middle.[12] Then, we solve the problem based on the refined grid, and continue with this procedure until the stopping criterion is met.

---

[12]This might not always be possible, e.g., if the resulting split would result in a non-integer number of new units. In such a case, we take the next smaller integer value.

3.4.4   Interpolation Methods

For the determination of the optimal decision in our approximated problem, we need an approximation of the value of the non-saved states. This approximation is obtained by interpolation. Different interpolation methods can be used. Grüne and Semmler [GS04] use a multilinear interpolation, while Munos and Moore [MM02] propose a simplex-based interpolation. In this subsection, we present these two methods. An empirical comparison is performed at the end of this chapter in Subsection 3.6.3.

3.4.4.1   Multilinear Interpolation

The multilinear interpolation uses all corner points of a hyperrectangle to interpolate the value of a point inside the hyperrectangle. Thereby, the value of a point in the space is interpolated by the corner values of the hyperrectangle in which the point is located. The following interpolation algorithm for a d-dimensional problem is taken from Davies [Dav97]:

- Pick an arbitrary axis. Project the query point along this axis to each of the two opposite faces of the box containing the query point.

- Use two $(d-1)$-dimensional multilinear interpolations over the 2^{d-1} datapoints on each of these two faces to calculate the values at both of these projected points.

- Linearly interpolate between the two values generated in step 2.

Assume that the considered hyperrectangle is the unit hyperrectangle with one corner at $(x_0, x_1, \ldots, x_{d-1}) = (0, 0, \ldots, 0)$ and the diagonally opposite corner at $(1, 1, \ldots, 1)$, then the weight w of each corner point ξ^i used for the interpolation of point x' is

$$w(\xi^i) = \prod_{k=0}^{d-1} \left(1 - |\xi_k^i - x_k'|\right),$$

where ξ_k^i is the $k-$th coordinate of ξ^i. The approximated value of a point x' inside the hyperrectangle with the corner points $\{\xi^0, \xi^1, \ldots, \xi^{2^d-1}\}$ is then obtained as:

$$\tilde{V}(x') = \sum_{i=0}^{2^d-1} w(\xi^i) \cdot V(\xi^i).$$

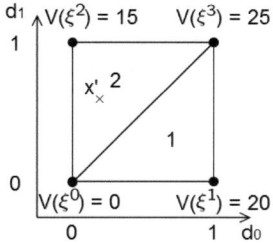

Figure 3.8: Example used to demonstrate multilinear and simplex-based interpolation. Depicted are the corner points of a hyperrectangle and the resulting triangulation.

Example (Multilinear interpolation). *Consider the example depicted in Figure 3.8. Assume that we want to interpolate the value of $x' = (0.2, 0.6)$.*
We calculate the weight of ξ^0 as $w(\xi^0) = (1 - |0 - 0.2|) \cdot (1 - |0 - 0.6|) = 0.32$. The weights for the other corner points are $w(\xi^1) = 0.08$, $w(\xi^2) = 0.48$ and $w(\xi^3) = 0.12$. Consequently, we approximate $\tilde{V}(x') = 0.32 \cdot 0 + 0.08 \cdot 20 + 0.48 \cdot 15 + 0.12 \cdot 25 = 11.8$.

Multilinear interpolation processes 2^d datapoints for the interpolation of a single value. For high dimensional problems, this becomes prohibitively expensive.

3.4.4.2 Simplex-Based Interpolation

In contrast to multilinear interpolation, simplex-based interpolation does not use all corners for the interpolation. Instead, it uses only $d + 1$ points. For this purpose, a hypercube is divided into $d!$ simplexes following a *Coxeter-Freudenthal-Kuhn triangulation* (Moore [Moo92]). Only the corner points of the simplex in which the interpolated state lies are used for the interpolation.
A *Coxeter-Freudenthal-Kuhn triangulation* can be obtained the following way: Assume that the considered hyperrectangle is the unit hyperrectangle. Then, the *Coxeter-Freudenthal-Kuhn triangulation* divides the hyperrectangle into $d!$ simplexes such that each of those simplexes corresponds to one possible permutation p of $(0, 1, \ldots, d - 1)$, and covers the region satisfying $1 \geq x_{p(0)} \geq \cdots \geq x_{p(d-1)} \geq 0$.

Example (Coxeter-Freudenthal-Kuhn triangulation). *Figure 3.8 shows a Coxeter-Freudenthal-Kuhn triangulation in the case of two dimensions. For two dimensions, there are two possible permutations $p_1 = (0, 1)$ and $p_2 = (1, 0)$. The first permutation results in simplex 1 satisfying $1 \geq x_0 \geq x_1 \geq 0$, the second permutation in simplex 2 with $1 \geq x_1 \geq x_0 \geq 0$.*

Note that we do not have to explicitly save all $d!$ simplexes of a hyperrectangle to apply the simplex-based interpolation. It is sufficient to determine the simplex in which a query point lies.

Munos and Moore [MM02] show how the simplex-based interpolation can be performed in $O(d \log d)$: Let $\{\xi^0, \ldots, \xi^{2^d-1}\}$ be the corners of the d-dimensional unit hyperrectangle. We choose the indices i of the corners such that $i = \sum_{k=0}^{d-1} \xi_{d-1-k}^i \cdot 2^k$, e.g., for a three-dimensional hyperrectangle, $\xi^0 = (0,0,0)$, $\xi^1 = (0,0,1)$, $\xi^2 = (0,1,0), \ldots, \xi^7 = (1,1,1)$.

To interpolate the value of a point x', we first have to find the simplex in which the point lies. For this purpose, we sort the coordinates of x' from the highest to the smallest, such that $1 \geq x_{j_0} \geq x_{j_1} \geq \cdots \geq x_{j_{d-1}} \geq 0$, where the indices $j_0, \ldots, j_{d-1}$ are a permutation of $\{0, \ldots, d-1\}$. Then, the indices $i_0, \ldots, i_d$ of the corners of the simplex containing x' are:

$$i_0 = 0, i_1 = i_0 + 2^{j_0}, \ldots, i_k = i_{k-1} + 2^{j_{k-1}}, \ldots, i_d = i_{d-1} + 2^{j_{d-1}} = 2^d - 1.$$

We can now use the barycentric coordinates of x' as weights of the corner points of the simplex used for the interpolation. Let $\xi^{i_0}, \ldots \xi^{i_d}$ be the vertices of the simplex in which x' lies. Then, the barycentric coordinates $\lambda_0, \ldots, \lambda_d$ of x' are defined as $\sum_{k=0}^{d} \lambda_k = 1$ and $\sum_{k=0}^{d} \lambda_k \cdot \xi^{i_k} = x'$. In the case of a Coxeter-Freudenthal-Kuhn triangulation, these barycentric coordinates are:

$$\lambda_0 = 1 - x_{j_0}, \lambda_1 = x_{j_0} - x_{j_1}, \ldots, \lambda_k = x_{j_{k-1}} - x_{j_k}, \ldots, \lambda_d = x_{j_{d-1}} - 0 = x_{j_{d-1}}.$$

With the barycentric coordinates, we can approximate the value of a point x' as

$$\tilde{V}(x') = \sum_{k=0}^{d} \lambda_k \cdot V(\xi^{i_k}).$$

The advantage of simplex-based interpolation is its efficiency. However, as it uses less points for the interpolation, its approximation quality tends to be worse than the quality of multilinear interpolation.

Example (Simplex-based interpolation). *Let us demonstrate how the value of point $x' = (0.2, 0.6)$ in Figure 3.8 is approximated using the simplex-based interpolation.*

First, we have to determine the indices j. Sorting the coordinates of x' in descending order leads to $j_0 = 1$ and $j_1 = 0$. Now, we can determine the indices i defining the corner points *of the simplex in which x' lies. We get $i_0 = 0$, $i_1 = 0 + 2^1 = 2$ and $i_2 = 2 + 2^0 = 3$. The simplex that contains x' is defined by the vertices ζ^0, ζ^2 and ζ^3.*

Next, we need to determine the barycentric coordinates. These are $\lambda_0 = 1 - 0.6 = 0.4$, $\lambda_1 = 0.6 - 0.2 = 0.4$ and $\lambda_2 = 0.2$. We approximate the value of x' as $\tilde{V}(x') = 0.4 \cdot 0 + 0.4 \cdot 15 + 0.2 \cdot 25 = 11$.

3.4.5 Splitting Criteria

The most important part of an adaptive aggregation method is the selection of regions that require a finer resolution. In our algorithm, the refinement is obtained by the splitting of the hyperrectangles that form the grid. The selection of the hyperrectangles that are split as well as the dimension, in which these hyperrectangles are split, is based on a splitting criterion.

In this subsection, we present different splitting criteria. These splitting criteria can be classified into three categories: criteria based on the value function, criteria based on the policy and a combination of both.

Splitting criteria based on the value function aim to obtain a good approximation of the value function. To this class of splitting criteria belong the *average corner-value difference* (ACVD) criterion and the *value non-linearity* (VNL) criterion proposed by Munos and Moore [MM02], as well as the rule proposed by Grüne and Semmler [GS04], to which we refer as *approximation error* (AE).

Policy based splitting criteria focus on the optimal policy to determine the splitting priority for a hyperrectangle. An example for a policy based criterion is the *policy disagreement* criterion proposed in [MM02]. It splits those hyperrectangles with a different optimal decision at the states that are the vertices of the hyperrectangle. However, as it performs relatively poor, we do not present this criterion in detail here. Instead, we present an alternative policy based criterion, which we call the *probability interpolated optimal solution* (PIOS) criterion.

The idea of a combination of a value function based splitting criterion and a policy based criterion is to focus on a good approximation of the value function around the (assumed) optimal policy. We present a combination of our *PIOS* criterion with different value function based criteria. A numerical evaluation of the different splitting criteria is made at the end of this chapter in Subsection 3.6.4.

In the following, we first describe the value function based criteria, before we introduce the PIOS criterion. Finally, we describe how these criteria can be combined.

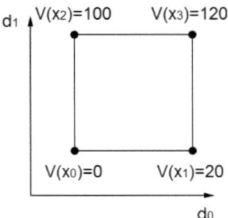

Figure 3.9: Two-dimensional rectangle used to illustrate the different splitting criteria.

3.4.5.1   Average Corner-Value Difference

The ACVD criterion is based on the idea that refinements are unnecessary if the value function is flat, i.e. values of neighboring states are similar.

To determine the splitting priority for a hyperrectangle, we calculate the average of the absolute difference of the values at the corners of the edges for all directions $i = 0, \ldots, d - 1$. The maximum of the calculated values corresponds to the splitting priority, the direction for which the maximum is taken is the splitting dimension.

Example. *Consider the two-dimensional example depicted in Figure 3.9. The splitting priority for dimension 0 is*

$$SP_0 = 0.5 \cdot (|V(x_0) - V(x_1)| + |V(x_2) - V(x_3)|) = 20,$$

while the splitting priority for dimension 1 is

$$SP_1 = 0.5 \cdot (|V(x_0) - V(x_2)| + |V(x_1) - V(x_3)|) = 100.$$

Consequently, the splitting priority for the rectangle is 100, and the splitting dimension is 1.

3.4.5.2   Value Non-Linearity

The ACVD criterion splits when the value function is not constant. However, this may result in many splits, even in cases when splits are not needed, because a good interpolation of the region is possible. The VNL criterion tries to avoid splits in such regions.

Instead of the absolute difference, the variance in the absolute increase of the values at the corners of the edges is calculated for all directions $0, \ldots, d-1$. The VNL criterion splits when the value function is not linear.

Example. *For the example shown in Figure 3.9, the variance in the absolute increase of the values at the corners of the edges is zero for direction 0, as the increase between x_0 and x_1 is the same as between x_2 and x_3. The same applies to the variance in direction 1. Hence, the splitting priority is 0 for the rectangle.*

3.4.5.3 Approximation Error

The ACVD criterion and the VNL criterion try to identify regions in the state space that should be refined based on the values of the corner points of a hyperrectangle. The advantage of this approach is that all required information is available, as the states at the corner of the hyperrectangle have already been evaluated during the previous optimization run.

A slightly different approach is taken by Grüne and Semmler [GS04]. They propose to refine all rectangles with an approximation error exceeding a predefined threshold value. To determine the approximation error, they use a set of test points x_i^t as depicted in Figure 3.10. For each of the test points, they calculate the approximation error. The splitting priority of a hyperrectangle is equal to the maximum approximation error.

Grüne and Semmler allow splits in several dimensions. For a better comparability with the other splitting criteria, we allow splits only in one dimension. We choose the splitting dimension such that the test point with the greatest approximation error lies on the hyperplane splitting the hyperrectangle.[13]

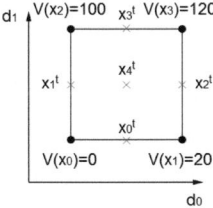

	$V(x)$	$\tilde{V}(x)$
x_0^t	10	10
x_1^t	55	50
x_2^t	80	70
x_3^t	130	110
x_4^t	65	60

Figure 3.10: Additional test points to estimate the approximation error.

Table 3.4: Evaluated values $V(x)$ and interpolated values $\tilde{V}(x)$.

---

[13]Grüne and Semmler use the same criterion, however, they split in all dimensions, for which the maximum approximation error of the test points lying on the corresponding hyperplane is relative large compared to the maximum approximation error of all points.

Example. *Consider the additional test points depicted in Figure 3.10. Table 3.4 shows the corresponding interpolated values,$\tilde{V}(x)$, and the real values, $V(x)$. The maximum approximation error is 20 and is made by the interpolation of x_3^t. Hence, the splitting priority of the rectangle is 20, and the splitting dimension is d_0.*

Note that for the calculation of the approximation error, the test points have to be evaluated (the optimal decision for the test points has to be determined). Compared to the other criteria, this method requires significantly more computational effort. The advantage of this method is that the error estimates are more appropriate than those obtained by the other methods.

3.4.5.4   Probability Interpolated Optimal Solution

The objective of the value function based splitting criteria is a good approximation of the value function. However, our true objective is to find the optimal policy for the considered problem. The previous splitting criteria may waste a significant amount of time refining regions that are not important because they are suboptimal. For example, it may never be profitable to build more than six units. In such a case, we do not need a very precise approximation of the value function in this region, even if the value function is highly non-linear there.

In the following, we first present the main ideas on which the new *PIOS* criterion is based. Then, we describe how the splitting priorities and the splitting dimension are determined.

Idea

Deviations between the optimal solution of the approximated problem and the optimal solution of the real problem can be caused by two issues: underestimation of the value of the optimal solution of the real problem, or overestimation of the quality of the optimal solution of the approximated problem.

As we do not know the optimal solution of the real problem, it is difficult to focus on a better approximation of the real solution. However, if we refine the grid such that all states of the optimal solution of the approximated problem are grid points, then we can exclude an overestimation of the optimal solution of the approximated problem.

A further argument to refine the state space around the optimal policy of the approximated problem is the following: if the value function of the considered problem is relatively smooth, then we can assume that states in the neighborhood of the optimal solution have also a high value function. Consequently, we might assume that the optimal solution of the real problem is somewhere in the neighborhood of the optimal solution of the approximated problem. Hence, it can be beneficial to concentrate the refinements in this area.

The PIOS criterion considers both of these ideas. Hyperrectangles that are used to interpolate the optimal solution of the approximated problem get the highest priority. Hyperrectangles that contain states of the optimal solution, but are not used to interpolate the optimal solution get a lower priority, while the priority of all other hyperrectangles is zero. To be able to prioritize the hyperrectangles correspondingly, we need to save the probability that a hyperrectangle is used to interpolate the optimal solution.[14]

Splitting Priority and Splitting Dimension

If a hyperrectangle is used to interpolate the optimal solution, the splitting priority of this hyperrectangle is set to one plus the probability of the optimal states that are interpolated by the hyperrectangle. If a hyperrectangle is not used to interpolate the optimal solution, but at least one of its corner points is part of the optimal solution, then the priority is equal to the probability of the optimal corner points of the hyperrectangle. Otherwise, the priority is set to zero.

Formally written, the splitting priority is

$$
SP = \begin{cases} 1 + \sum\limits_{x^* \in \mathcal{X}^{\text{Rect,I}} \cap \mathcal{X}^{\text{Opt}}} P(x^*) & \text{if } \mathcal{X}^{\text{Rect,I}} \cap \mathcal{X}^{\text{Opt}} \neq \emptyset, \\ \sum\limits_{x^* \in \mathcal{X}^{\text{Rect,C}} \cap \mathcal{X}^{\text{Opt}}} P(x^*) & \text{otherwise.} \end{cases}
\tag{3.9}
$$

Thereby, $\mathcal{X}^{\text{Rect,I}}$ denotes the set of all states whose value is interpolated using the considered hyperrectangle. These are all states that lie within the hyperrectangle or at the border of the hyperrectangle, excluding the corners of the hyperrectangle. The set of states at the corners of the hyperrectangle is $\mathcal{X}^{\text{Rect,C}}$, while the set of states belonging to the optimal solution is $\mathcal{X}^{\text{Opt}}$. The probability that a state is visited when the optimal policy is applied is $P(x^*)$.

Note that formula (3.9) only calculates the splitting priority of a hyperrectangle, but not the splitting dimension. To determine the splitting dimension, we use the *VNL* criterion.

---

[14]This criterion is only suitable for stochastic problems. Otherwise, only one state per stage belongs to the optimal solution. Consequently, only one rectangle per stage would be refined per iteration.

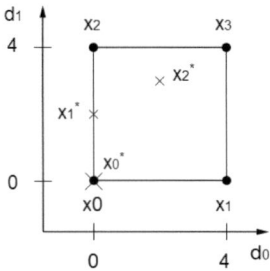

	Coordinates	$P(x)$
x_0^*	(0,0)	0.15
x_1^*	(0,2)	0.15
x_2^*	(2,3)	0.05

Figure 3.11: Rectangle including points of the optimal solution.

Table 3.5: Points belonging to the optimal solution.

Example. *Consider the example depicted in Figure 3.11. Three points (x_0^*, x_1^*, x_2^*) of the optimal solution lie at the edge or inside the rectangle. The coordinates and the probabilities of the corresponding points are given in Table 3.5.*

Note that point x_0^* does not need to be interpolated, as it belongs to the grid. Hence, x_0^* is not considered to determine the splitting priorities. Consequently, the splitting priority for the rectangle is equal to 1.2, the sum of the probabilities for x_1^* and x_2^* plus 1.

As we have seen before, the splitting priorities of the VNL criterion are 0 for both dimensions. In such a case, we select the first possible splitting dimension, in this example dimension 0.

3.4.5.5   Combined Splitting Criteria

The splitting criteria presented so far are either value function based criteria or policy based criteria. On the one hand, value function based criteria deliver a good approximation of the value function, but may waste significant effort in refining non-optimal regions. On the other hand, the PIOS criterion delivers a realistic approximation of the optimal solution of the approximated problem. However, it may underestimate the optimal solution of the real problem. Due to this underestimation it may not refine the grid in this region.

A combination of a value function based and a policy based splitting criterion might overcome such problems. We can combine any of the value function based criteria with the PIOS criterion. As the approach is the same for all combinations, we first describe how we combine the criteria. Then, we give an example for each of the combined criteria.

There are several possibilities to determine the splitting priority when different criteria are combined. One possibility is to use a weighted sum of the splitting priorities of both criteria. However, the problem is that the priorities calculated by the different splitting criteria differ significantly. While the priorities calculated by the PIOS criterion are in the interval between 0 and 2, those of the ACVD for example can be much greater. To make the priorities comparable, they must be normalized.

To avoid this problem, we might use the product of the splitting priority of both splitting criteria as priority for the hyperrectangle. However, also the product of both priorities is problematic. The reason is the way we defined the PIOS criterion. If a hyperrectangle is used to interpolate the value of a state of the optimal solution, the splitting priority for this hyperrectangle is between 1 and 2. Consequently, if there are two hyperrectangles A and B that are used to interpolate the optimal solution, the priority of A can be at most twice the priority of B. For any of the value function based criteria, the priority of hyperrectangle A can be much more important than the one of B (or the other way round). Hence, if we use the product of both splitting priorities, much more weight is given to the value function based criterion.

To get a more balanced result, we define the splitting priority for the combined criteria as follows:

$$SP = \begin{cases} SP^{\text{VF}} \cdot (SP^{\text{PIOS}} - 1) + 2 & \text{if } SP^{\text{PIOS}} > 1, \\ \min(2, SP^{\text{VF}} \cdot SP^{\text{PIOS}} \cdot \epsilon_1 + 1) & \text{if } 0 < SP^{\text{PIOS}} <= 1, \\ \min(1, SP^{\text{VF}} \cdot \epsilon_2) & \text{otherwise.} \end{cases}$$

Thereby, SP^{VF} denotes the splitting priority of the value function based splitting criterion, SP^{PIOS} the one calculated by the PIOS criterion.

We define three different intervals for the splitting priority: if a hyperrectangle is used to interpolate the value of a state of the optimal solution, then we ensure that the splitting priority is at least 2.[15] Note that the value of $SP^{\text{PIOS}} - 1$ is in the interval between 0 and 1. Consequently, the importance of a hyperrectangle relative to another hyperrectangle is no longer restricted to a maximum of 2.

If a hyperrectangle is not used to interpolate the optimal solution, but at least one of the states at its corners belongs to the optimal solution, we assign a priority between 1 and 2 to that hyperrectangle. To achieve this, we multiply the product of both splitting priorities with a small constant ϵ_1 and add 1 to the result. The

---

[15]The value of 2 is chosen arbitrarily. If we split a fixed number of hyperrectangles per iteration, we are only interested in the relative priorities of the hyperrectangles. Only if we use a variable number of splits per iteration, the (absolute) splitting priority of a hyperrectangle is important.

constant should be chosen such that the product of the splitting priorities and the constant is less than 2. For example, we might choose $\epsilon_1 = 0.0001$.

If the splitting priority of the PIOS criterion is equal to zero, the splitting priority for the hyperrectangle is 1 or less. In such a case, we use the priority calculated by the value function based criterion and multiply it with another small constant ϵ_2. For the determination of the splitting dimension, we use the value function based criterion.

Let us briefly demonstrate this approach based on the examples already used for the different criteria. Thereby, we refer to the combination of the PIOS with the AE criterion as *weighted approximation error* (WAE). The other two combinations are correspondingly named *weighted average corner-value difference* (WACVD) and *weighted value non-linearity* (WVNL). As the splitting dimension corresponds to the splitting dimension selected using the respective value function based criterion, we omit the splitting dimension in the following examples.

Example (Weighted approximation error). *Let us consider the example depicted in figures 3.10 and 3.11. The splitting priority of the AE criterion is 20, the splitting priority of the PIOS criterion is 1.2. Consequently, the splitting priority for the rectangle is* $SP = 20 \cdot (1.2 - 1) + 2 = 6$.

Example (Weighted average corner-value difference). *Let us now consider the example shown in figures 3.9 and 3.11. The splitting priority of the ACVD criterion is 100, the splitting priority of the PIOS criterion is 1.2. Hence, the splitting priority for the rectangle is 22.*

Example (Weighted value non-linearity). *The splitting priority of the VNL criterion based on the example depicted in Figure 3.9 is 0. As the splitting priority of the PIOS criterion is greater than 1, the splitting priority is 2.*

3.5   Algorithm

In this section, we give an overview of the different parts of the algorithm that we use to solve the power generation expansion problem. We start with a description of the algorithm. Next, we go in details about how we select the initial grid used in the algorithm. Finally, we conclude this section with some remarks on the implementation.

Algorithm 4 ISDP algorithm to optimize power plant investments

---

1: Simulate fuel prices
2: Select initial grid
3: Determine transition probabilities
4: Select representative fuel price scenarios for the saved states
5: Calculate the expected contribution margins for the saved states
6: Solve the approximated problem
7: **while** Stopping criterion is not met **do**
8:     Calculate splitting priorities and splitting dimension for each hyperrectangle
9:     Refine the grid based on the calculated splitting priorities
10:     Calculate the (expected) contribution margins for the new states
11:     Solve the refined approximated problem
12: **end while**

3.5.1 Overview

Algorithm 4 gives a pseudo-code description of the approach we use to solve the power generation expansion problem. The first step is the simulation of fuel prices. We assume that the stochastic process used to simulate the prices is Markovian. Then, based on these simulations, we select the states of the initial grid as described in the next subsection. Next, we use the simulated fuel prices to determine the transition probabilities between the states of succeeding stages. Then, we select the representative fuel price scenarios for each of the saved states.

Afterwards, we determine the expected contribution margin for every saved state. This contribution margin is based on the optimization of the power generation with one of the fundamental electricity market models described in Chapter 2. For this purpose, we use the representative fuel price scenarios of the considered state. The expected contribution margins for the non-selected fuel price scenarios are interpolated. Next, we use the backward induction to solve the initial approximated SDP problem.

Then, we enter a loop in which the approximated problem is iteratively refined and solved again. To refine the problem, we calculate the splitting priority and the splitting dimension for all hyperrectangles of the grid. For each stage of the problem we select some—either a fixed or a variable number of—hyperrectangles that are split. We add the states at the corner points of the split hyperrectangles to the approximated problem. Next, we determine the expected contribution margins for these states. Then, we solve the refined problem. We continue with this loop, until the stopping criterion is met. This stopping criterion may be a fixed

number of iterations. Alternatively, we may stop when the optimal solution did
not change for a predefined number of iterations, or if the expected value of the
optimal solution does not change any more.

3.5.2 Choice of the Initial Grid

An issue we have not discussed so far is the choice of the initial grid, defining
which states are used for the initial approximation. Two decisions have to be
taken: the choice of the distance between two neighboring grid points in each
dimension, and the minimal and maximal values of the grid points in each dimension.

The more important decision is the second one. While the number of states saved
between the extreme points is adapted during the optimization (and therefore also
the distance between the grid points), the extreme points of the hyperrectangle
are fixed. Solutions outside of the hyperrectangle cannot be reached during the
optimization.

Remember that a state in our model consists of several attributes, which can be
grouped into two categories: attributes representing power plants, and attributes
representing fuel prices. For each of these attributes, we define a set of initial
values. In the following, we first describe how we select the initial values for the
attributes representing the new units, before we discuss our choice of the initial
values for attributes representing the fuel price scenarios. The initial states are
obtained as the set of all possible permutations of the values of the attributes.

3.5.2.1 Units

The choice of the initial values for attributes representing power plants is straightforward. We define a maximum number of allowed new units for each unit type,
$x^{\max,u}$. Based on the idea to limit market power, the number of allowed new units
of the considered operator may be chosen such that a predefined market share
cannot be exceeded. The number of allowed new units may depend on the considered operator. For a new entrant to the market, we may allow more new units
compared to one of the big four operators (E.On, RWE, EnBW, Vattenfall). The
minimum number of new units is obviously zero.

The maximum number of new units that can be built by external competitors
should be chosen such that it does not restrict the investment decisions of the

external investment module. This number depends amongst others on the optimization horizon and the chosen investment module. If external investments are only based on the scarcity of supply, less investments are carried out by competitors compared to the case when additional investments are done based on the expected profitability.

Besides the minimum and maximum number of new units, we specify the number of scenarios for each unit that should be initially saved. This number must be between 2 and $x^{\max,u} + 1$.[16] Then, we select the initial values for these attributes such that all neighboring values are equidistant. If equidistant values result in non-integer values, we use the next lower integer instead. For example, if we allow nine new CCGT units and start with three different scenarios, we take 0, 4 and 9 as initial values.

3.5.2.2 Fuel Prices

One might be tempted to use the same approach for attributes representing fuel prices. There are two problems though. First, it might be difficult to define a pertinent upper bound (and if the climate spread is used, this applies also to the lower bound) for fuel prices a priori. Second, the choice of equidistant fuel prices is not necessarily reasonable, as fuel prices have different occurrence probabilities.

Hence, we propose an alternative approach, taking into account these difficulties. We start by specifying a confidence interval for fuel prices. Based on the fuel price simulations carried out before, we determine minimum and maximum fuel prices such that a predefined percentage of the fuel prices lies within this interval. For example, assume that we use a 95% confidence interval. In that case, the 2.5 percentile of the simulated fuel prices forms the lower bound, while the 97.5 percentile corresponds to the upper bound.

Then, we specify the number of initial fuel price scenarios. To obtain the additional values between the maximum and the minimum fuel price scenarios, we use equidistant percentiles instead of equidistant values. For example, if we decide to start with five fuel price values and a 95% confidence interval, then the initial fuel prices are those corresponding to the 2.5, 26.25, 50, 73.75 and 97.5 percentiles.

---

[16]If the number was one, the investment decision would be fixed. Therefore, we do not allow this case.

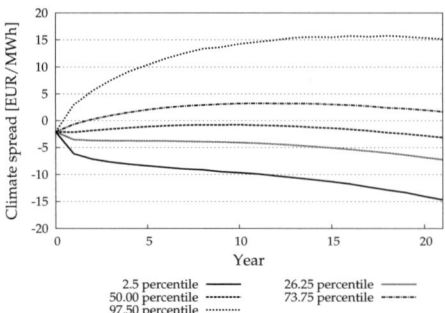

Figure 3.12: Example for climate spread percentiles used as initial values. Logs of fuel prices are simulated as trended mean-reversion process.

Figure 3.12 illustrates the previously mentioned example based on 50,000 fuel price simulations, where the logs of fuel prices are simulated as trended mean-reversion process. Two advantages of this approach can be seen in the figure: First, the resolution around most probable fuel prices is higher than the one at the tail of the distribution. Second, the intervals adapt to increasing uncertainty. For later stages, uncertainty increases and thereby also the considered interval increases. To take this increasing uncertainty into account, one may decide to use more initial fuel price scenarios at later stages. This is especially beneficial if fuel prices are simulated as a GBM, where the interval between maximum and minimum climate spread can get significantly larger than the one shown in Figure 3.12.

3.5.3   Remarks on the Implementation

Let us briefly comment on two details of the implementation: parallelization and efficient data structures for the interpolation.

3.5.3.1   Parallelization

A challenge of an SDP formulation of the power generation expansion problem is the huge state space. For every state, the contribution margin must be calculated. Especially if an LP-based fundamental electricity market model is used, these computations are very time intensive.

Since each run of the fundamental electricity market model is independent of the others, the calculation can be completely computed in parallel. Concerning the solution of the ISDP problem, we can parallelize the determination of the optimal solution for states of the same stage. In addition, the calculation of splitting priorities for all hyperrectangles of the grid can be performed in parallel.

3.5.3.2 Efficient Data-Structures for the Interpolation

For the interpolation of states that are not saved, we first have to find the hyperrectangle in which the requested state is saved. To allow an efficient retrieval of specific hyperrectangles, we follow the proposition of Munos and Moore [MM02] and use a kd-trie, a special case of a kd-tree, to store the hyperrectangles.

A kd-tree is a binary tree, in which every node represents a k-dimensional point, or, as in our case, a k-dimensional hyperrectangle. Every (non-leaf) node generates a splitting hyperplane that divides the space into two subspaces. Points or hyperrectangles left of the hyperplane represent the left subtree, points or hyperrectangles right of the splitting hyperplane represent the right subtree. While the splitting hyperplane of a kd-tree can be somewhere between the lower and the upper value of the hyperrectangle, a kd-trie splits always in the center. For a detailed description of kd-trees, see, e.g., Moore [Moo91].

In our case, the root of the kd-trie is the hyperrectangle covering the whole state space. All other nodes are hyperrectangles covering only a subset of the state space. The leafs of the tree represent the "active" hyperrectangles used for the interpolation of non-saved states. Every time a hyperrectangle is split, the two new hyperrectangles are added as left and right subtree of the hyperrectangle that is split.

3.6 Evaluation

In this section, we evaluate our algorithm according to several criteria. We start with an outline of how the tests were performed and a description of the test data. Then, we test the influence of different parameters for the algorithm. Finally, we compare the performance of the algorithm with alternative approaches.

A first parameter test compares the approach based on an explicit representation of fuel and carbon prices with the representation of the climate spread. While the representation based on the climate spread is faster and allows to model other parts of the model in more detail, the quality of the solutions is only slightly worse than those based on an explicit fuel price representation.

Next, we compare multilinear interpolation with simplex-based interpolation for the approximation of the value function of non-saved states. It turns out that multilinear interpolation works better for our problem.

Afterwards, we compare the different splitting criteria that have been presented in Subsection 3.4.5. According to our evaluations, the splitting criteria based on a combination of a value function based criterion with the PIOS perform best. In general, the WAE criterion works best, despite the additional computational burden caused by the calculation of the approximation errors.

Besides the choice of the splitting criterion, we also test whether a fixed number of splits or a variable number of splits performs better. The results indicate that a variable number of splits yields better performance.

In the last part of this section, we compare the quality of the solutions obtained by our algorithm with the quality of solutions obtained by alternative methods. First, we compare our algorithm to a deterministic approach using the expected value for the fuel prices. Our algorithm outperforms the deterministic approach, since the latter one disregards the stochasticity of fuel prices.

Finally, we compare the solution of our ISDP approach to the solutions of a normal SDP approach. The main advantage of our algorithm is its capability to handle more complicated problems. For comparison reasons, we consider lower dimensional problems that can also be solved by a normal SDP approach. Even for these simpler problems, it is still preferable to apply our algorithm under certain conditions.

3.6.1   Test Data

For all the tests in this section, we use our LDC model with scarcity mark-ups to determine the contribution margins for the states. We perform tests with different parameters like the number of new units, the optimization horizon, the depreciation method or the stochastic process used for the fuel price simulations.

More important than the exact test configuration is in our opinion to understand the way we evaluate our tests. During each iteration of our algorithm, we solve an approximated problem to optimality. Due to the approximations, the expected value of the optimal solution of the approximated problem does in general not correspond to the real value of this solution.

To measure the quality of the solution of the approximated problem, we evaluate this solution based on a Monte Carlo simulation of fuel prices. After each iteration of our algorithm, we simulate 10,000 fuel price scenarios for each year of the

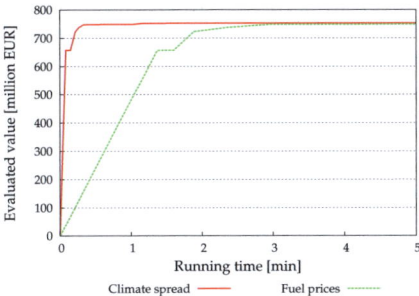

Figure 3.13: Solution quality based on climate spread and explicit fuel price representation.

optimization horizon. Based on these simulations, we determine the profits we get if the optimal solution of the approximated problem is applied. To keep the evaluation results comparable, we use the same 10,000 fuel price scenarios for the evaluation of all solutions. Note however, that these fuel price scenarios are not the same as those used within the optimization. For the comparison of the interpolation method as well as for the comparison of the different splitting criteria, we split ten hyperrectangles per stage and iteration.

3.6.2 Fuel Price Aggregation

One of the techniques we have proposed to reduce the dimensionality of the problem is to replace the explicit representation of fuel prices and the carbon price with an aggregated one, based on the climate spread. In this subsection, we briefly test the effects of this aggregation. To be able to consider problems with an explicit representation of fuel and carbon prices, we have to neglect investments by competitors to limit the size of the state space.

An exemplary comparison of both approaches is shown in Figure 3.13. Due to the significant smaller state space, the approach based on the climate spread runs faster, and hence it finds a good solution earlier. Once both algorithms are converged, the solution of both approaches are of very similar quality. The approach based on an explicit fuel price representation performs only slightly better. In almost all of our tests, differences between the evaluated solutions are below 1%. As the climate spread approach allows us to model other characteristics of the problem, e.g., the representation of investments by competitors, in more detail without significantly degreasing solution quality, we opt to use the climate spread

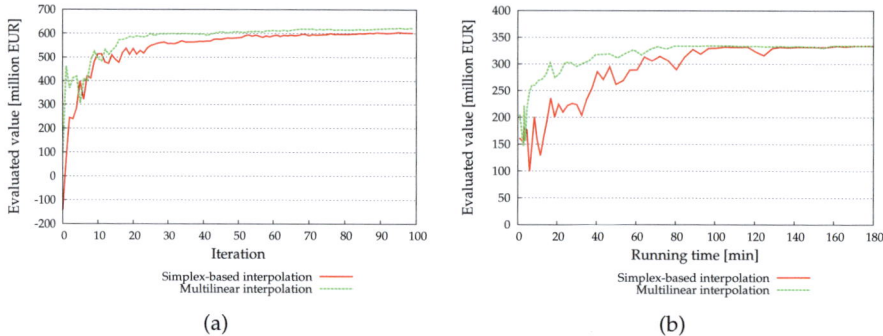

Figure 3.14: Comparison of the solution quality obtained with simplex-based interpolation and multilinear interpolation for two different data sets.

representation for the following comparisons. However, it should be remarked that this approach is especially suited to compare coal units and CCGT units. If additional units, e.g., lignite units, should be imcluded in the comparsion, the approach might perform worse.

3.6.3   Simplex-Based Interpolation vs. Multilinear Interpolation

We have presented two methods for the interpolation of the value of non-saved states: simplex-based interpolation and multilinear interpolation. Simplex-based interpolation uses less points for the interpolation than multilinear interpolation. Consequently, simplex-based interpolation is faster than multilinear interpolation, but in general less accurate.

Both methods have been compared by Davies [Dav97] based on two different problems: a two-dimensional and a four-dimensional problem. Davies reports that multilinear interpolation needs (especially in the four-dimensional case) less attempts to find the optimal solution, but simplex-based interpolation is up to two times faster.

Our observations differ. Concerning the number of iterations before the algorithm converges, we observe a similar result as Davies. The application of multilinear interpolation results in faster convergence in terms of iterations than simplex-based interpolation. This behavior can be explained by the more accurate approximation results of multilinear interpolation.

Figure 3.14(a) shows an exemplary comparison of both approaches. It depicts the quality of the obtained solutions using multilinear and simplex-based interpolation as a function of the number of iterations of the algorithm. The comparison is based on a five dimensional problem. Especially during the first iterations, when the number of saved states is small and many values have to be interpolated, the more accurate multilinear interpolation leads to a better solution.

Concerning the running time of the algorithm, our observations differ from those reported by Davies. While simplex-based interpolation is slightly faster per iteration, this computational advantage does not outweigh the slower convergence, as it can be seen in Figure 3.14(b), showing the quality of obtained solutions as a function of the running time for another five-dimensional problem.

Let us briefly add two comments, which are independent of the interpolation method. In Figure 3.14, we see two typical patterns for our algorithm. First, the greatest improvements of the solutions are obtained during the first iterations. Afterwards, the solution may still improve, but only to a smaller extent. Second, a refinement does not automatically lead to a better solution. Especially during the first iterations, refinements may result in a considerably worse solution.

3.6.4 Comparison of Splitting Criteria

Probably the most important design parameter of our algorithm is the splitting criterion. A good splitting criterion should be reliable and efficient Grüne and Semmler [GS04]. A reliable splitting criterion is able to detect hyperrectangles with great approximation errors. An efficient splitting criterion avoids unnecessary refinements. As we cannot directly measure the efficiency and reliability of a splitting criterion, we compare the different splitting criteria based on numerical tests with different problem parameters.

The importance of the splitting criterion depends on the problem parameters. Some data sets turn out to be relatively easily solvable. These are mostly problems that do not consider investments by competitors, or that are exposed to a low degree of fuel price uncertainty. Here, all splitting criteria work relatively well.

Figure 3.15(a) shows an example of such an problem without external investments. The performance of the different splitting criteria can hardly be distinguished. If anything can be noticed, then that the value function based splitting criteria perform slightly worse than the other ones. Amongst the value function based criteria, the AE criterion performs worst in this case. This is an untypical

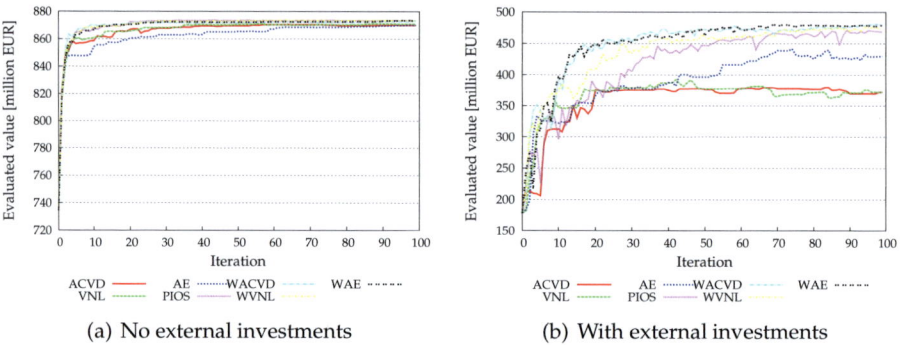

(a) No external investments   (b) With external investments

Figure 3.15: Solution quality for different splitting criteria for a problem without external investments (a) and (a different problem) with external investments (b).

result—in most of our tests, the AE criterion performs best amongst the different value function based criteria.

There are other problems though, where the choice of the splitting criterion is more important. Consider the results depicted in Figure 3.15(b). The data set used for this problem is similar to the one before. The difference is that we consider now investments of competitors.

Comparing both figures, two things stand out: First, the convergence of the algorithm is much slower. While the solutions in Figure 3.15(a) improve only slightly after the first ten iterations, it takes much longer until the algorithm converges for the problem depicted in Figure 3.15(b).

Second, the performance of the splitting criterion plays a more important role for the second data set. The value function based criteria perform significantly worse than the other criteria, especially the ACVD criterion and the VNL criterion. The PIOS criterion performs similar to the value function based criteria during the first 30 iterations. Then, it starts to outperform the value function based criteria. The combined criteria work best in this case. The WAE and the WACVD perform best in terms of number of iterations.

The results of our tests indicate that there is no single criterion that is best in any case. In most cases, the combined criteria perform best, however it is hard to say which one of the combined criteria should be used. In general, the WAE results in the fastest convergence in terms of iterations. The calculation of the approximation error can take a long time though. Sometimes, the more accurate estimate of the approximation error outweighs this additional time, sometimes not.

Comparing the PIOS criterion with the value function based criteria, results are ambiguous. In the majority of the test cases, the PIOS criterion performed better, but we observed also the opposite.

A difference between both kinds of criteria can be seen regarding Figure 3.15(b). The curve of the PIOS criterion exhibits more ups and downs compared to the value function based splitting criteria. The PIOS criterion focuses on an adequate estimate of the value of the optimal solution for the approximated problem. Possible overestimations are soon corrected, while the value function approximations of the other states are less accurate. Consequently, the algorithm may select another overestimated solution that is in reality worse than the current solution. After this overestimation is corrected, the algorithm proceeds to the next solution. The value function based criteria try to correct approximation errors in the whole state space. As a consequence, the approximation error of the optimal solution of the approximated problem is in general similar to the approximation error in other regions of the state space. Hence, value function based criteria change the optimal solution less frequent due to an overestimation somewhere else in the state space. The different approaches of the splitting criteria can also be seen in Figure 3.16. This figure shows the difference between the expected value of a solution and the evaluated value of the solution (for the same data set as the one used in Figure 3.15(b)). During the first iterations, the solution quality is significantly overestimated.[17] Splitting criteria considering the optimal policy correct this overestimation much faster than purely value function-based criteria. It is interesting to see that the combined criteria correct this overestimation slightly faster than the PIOS criterion.

3.6.5 Fixed Number of Splits vs. Variable Number of Splits

Besides the choice of the splitting criterion, we also have to select the number of rectangles that are split. Basically, we choose between two options: a fixed number of splits or a variable number of splits per iteration and stage.

We compare the performance of a fixed number and a variable number of splits based on the WACVD criterion and the WAE criterion. We choose these two criteria, as those are the criteria that performed best on average in our previous tests. For the evaluation of a fixed number of splits, we decided on the ten rectangles per stage and iteration with the highest splitting priority. We exclusively select

---

[17]This is not always the case. It depends on the chosen parameters (e.g., initial grid) whether solution quality is overestimated or underestimated.

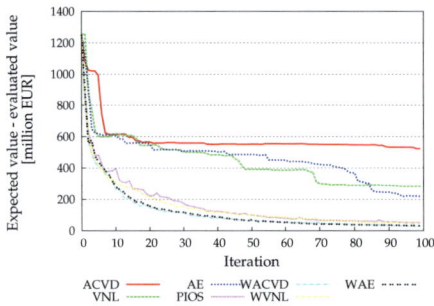 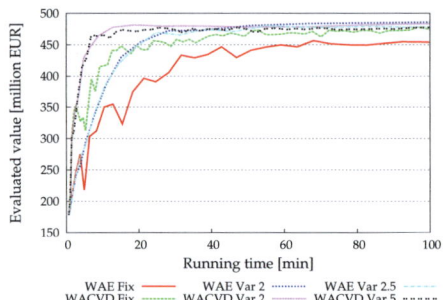

Figure 3.16: Difference between the expected value of the optimal solution of an approximated problem and the evaluated quality of that solution.

Figure 3.17: Comparison of a fixed number and a variable number of splits per iteration.

hyperrectangles with a splitting priority greater than zero. If there are less than ten rectangles with a splitting priority greater than zero, we only split the rectangles with a positive priority.

For the application of a variable number of splits, all rectangles with a splitting priority greater than a predefined threshold value are split. If there are less than ten hyperrectangles exceeding the threshold value, we split the ten hyperrectangles with the highest probability greater than zero.

For this comparison, we use two different threshold values. The first one is a threshold value of two. This relatively low threshold value results in many splits per iteration. Every hyperrectangle that is used to interpolate the optimal solution is split using this threshold value. Note that applying a threshold value of two results in (almost) the same splits for both different splitting criteria.[18]

The second threshold value is different for both criteria. For the WAE criterion, we choose a threshold of 2.5. This threshold value leads to splits for all hyperrectangles, where the product of the probability that they are used to interpolate the optimal solution and the approximation error is greater than 0.5 million €. For the WAE, we use a threshold value of 4. This higher threshold value is motivated by the fact that the corner-value difference is in general (much) higher than the interpolation error.

The results of an exemplary comparison are shown in Figure 3.17. The quality of the optimal solutions is depicted as a function of the running time. For this

---

[18]If there are less than ten hyperrectangles with a splitting priority greater than two, the splits may be different.

problem, the WACVD performs better than the WAE criterion. For both splitting criteria, the fixed number of splits performs worst. The variable number of splits show a similar performance for the different threshold values, with a small advantage for the lower threshold value.

However, there is a drawback of a low threshold value for the variable number of splits that cannot be seen in Figure 3.17: it may add too many states to the approximated problem. After one hour of running time, the number of considered states is about twice (for the WAE) respectively four times (for the ACVD) as high as for the splitting criterion with a threshold value of two compared to the splitting criterion with the higher threshold value. If we use a more time-intensive fundamental electricity market model, the evaluation of the additional states might change the advantageousness of the different threshold values.

3.6.6 Comparison with Alternative Approaches

The evaluations we presented so far focused on finding the best parameters for our algorithm. The results we have seen indicate that our algorithm works. In most cases, the quality of the obtained solutions improves during the runtime of the algorithm. However, there are some results questioning the performance of the algorithm. For some splitting criteria, the algorithm seems to be converged, but not to the optimal solution. This leads to the question whether the results of the algorithm using the best splitting criterion are (nearly) optimal, or whether there may exist a much better solution.

Unfortunately, this question cannot be answered. Due to the stochasticity of the problem and the huge state space, we cannot calculate the optimal solution. To get an idea how good the solution of our algorithm is, we compare the quality of the solutions of our algorithm applying the WAE criterion to the quality of solutions of two alternative approaches. The first method we use for the comparison is a deterministic DP problem based on expected fuel prices. The second method is a normal SDP approach without interpolation.

3.6.6.1 Expected Value Approach

A simple method to reduce the complexity of the power generation expansion problem is to disregard the stochasticity of fuel and carbon prices. Omitting the uncertainty reduces the state space significantly, such that the simplified problem can be solved to optimality. In the following, we first describe how we model the

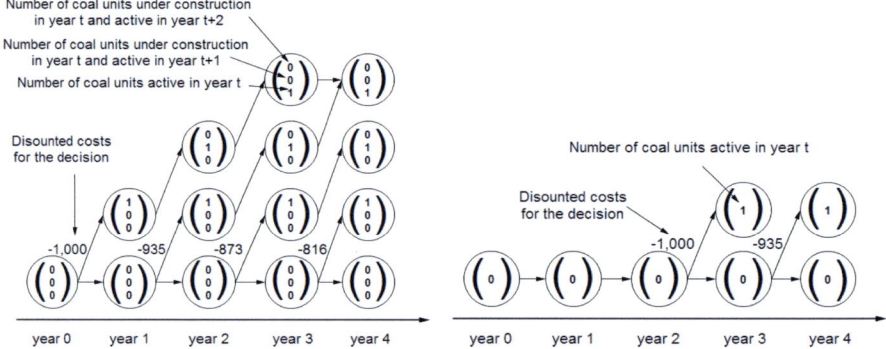

(a) Example with additional attributes to consider construction time.
(b) Example without additional attributes to
consider construction time.

Figure 3.18: Two approaches to model a deterministic power generation expansion problem.

deterministic problem, before we compare the results of the deterministic formulation with the results of our algorithm.

Description

The determination of the deterministic problem is straightforward. We replace the different fuel price scenarios for every year with the expected value of the fuel prices. In addition, we allow investments in new power plants every year.

In a complete deterministic setting, we do not have to add extra attributes for the construction time. Instead, we assume that new units are immediately available the next year. To consider construction times, we must shift back the decisions by the construction time minus one year. Consequently, the first year in which we allow new decisions is the year equal to the construction time minus one. Let us illustrate this approach with a simple example.

Example. *Assume that we might invest in a coal unit. The construction time of the unit is three years. In a stochastic setting, a state uses three attributes to represent the number of units: an attribute for the number of coal units active at the moment, an attribute for the number of coal units under construction that are finished the next year, and an attribute for the number of coal units under construction that are finished in two years. Figure 3.18(a) illustrates this problem formulation. Assume that the investment costs for a new unit are 1,000 million €. With an annual discount rate of 7%, we obtain the discounted investment costs shown in the figure.*

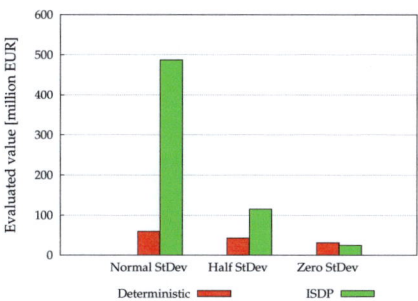

Figure 3.19: Comparison between the expected value approach and our algorithm for different degrees of uncertainty

Figure 3.18(b) depicts the equivalent problem without additional attributes for the number of units under construction. Two things should be noted: First, the discount factor for the decision costs must be applied with care. As the decision taken during the second year in Figure 3.18(b) is in reality taken during the initial year, we have to apply the discount factor associated with the initial year (and not the one associated with the second year).

The second thing to note is that such a representation is only possible in a deterministic case. In a stochastic setting, we get additional information (e.g., about the fuel price development) as time goes by. One may use this additional information to improve his decision. In the deterministic case, no additional information is obtained. Hence, the solutions of both models models shown in Figure 3.18 are the same.

Comparison

Figure 3.19 compares the evaluated value of the optimal solution of the deterministic approach with the evaluated value of the solution of our ISDP approach. The tests are based on a simulation of the logarithm of fuel prices as correlated GBMs. To demonstrate the effect of uncertainty, we vary the standard deviation of the GBMs. As a basis, we use a standard deviation of 0.23 for the logarithm of the annual gas price change and 0.14 for the logarithm of the annual coal price change. These values are based on the annual import prices of fuels in Germany [BMW09] deflated using the consumer price index. We compare the quality of the solutions based on fuel price simulations with these standard deviations, with half of these standard deviations and with a standard deviation of zero, i.e., without uncertainty.

In both test runs with stochastic prices, our model performs significantly better than the expected value approach. The higher the uncertainty, the greater the ad-

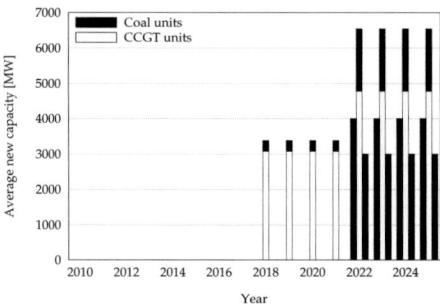

Figure 3.20: Optimal solution of the deterministic approach and the ISDP model. First column shows shows optimal solution of the deterministic model, second column the average values of the optimal ISDP solution, third column is the optimal ISDP solution for the expected fuel prices.

vantage of an approach explicitly accounting for it. Only in the test run without uncertainty, the deterministic model outperforms our model. This is possible because we allow annual decisions in the deterministic model, while in our model, decisions are taken every four years.

The optimal solution for the data set with the normal standard deviation is shown in Figure 3.20. We use three different columns to depict the optimal deterministic solution, the average values for the optimal solution of our approach and the optimal solution of our approach for the expected fuel price scenarios.

In the first column, the optimal solution of the expected value approach is shown. The optimal solution of this approach consists only of some coal units built at the end of the optimization horizon. The second column depicts the average new capacity built following the optimal solution of our approach. As the optimal solution depends on the realized fuel prices, there is not a single optimal solution per year. The depicted values are the average values of new units built during the evaluation based on 10,000 fuel price simulations. The third column shows the optimal solution of our algorithm for the expected fuel prices.

Comparing the solution of our algorithm with the expected value approach, two observations are remarkable. First, the optimal solution of our algorithm starts earlier with adding new capacity, and it also builds more new units on average. As our algorithm considers the stochasticity of fuel prices, it starts building new units if fuel prices develop such that one technology is more profitable than the other one. However, if the fuel price development follows the expected values, our algorithm comes to a similar solution as the expected value approach.

Secondarily, it is remarkable that our algorithm prefers CCGT units. While the expected value approach builds only coal units, our algorithm builds more CCGT units on average. There are two reasons for the construction of CCGT units. First, they are built when fuel price development favors CCGT units. Second, they are built because they are cheaper than coal units. Consequently, the risk is lower as they require fewer years for amortization. These observations are in line with the results by Dixit and Pindyck [DP94] that uncertainty favors less capital intensive investments.

To be fair, it has to be mentioned that the value of a stochastic model is in general lower than the difference between both approaches. The comparison is based on the assumption that the planning process is carried out only once. In reality, several re-planning steps will take place. During this re-planning, expected fuel prices can be adjusted to the observed fuel price development. This rolling horizon approach generally leads to better results compared to one single planning. Therefore, the advantage of a stochastic approach is in general smaller than the one we observe here.

3.6.6.2 Stochastic Dynamic Programming Approach

While the comparison with the expected value approach shows that our algorithm performs better, this result is not surprising. Considering a stochastic problem, any stochastic approach should outperform a deterministic one.

A more meaningful comparison is one with other stochastic methods. A common approach is normal SDP (e.g., Jensen and Meibom [JM08], Klein et al. [Kle+08]). Note that our problem formulation is also an SDP formulation. The difference between our approach and a normal SDP formulation is the interpolation and refinements we use. As it is computationally intractable to solve the problem in full resolution, our algorithm tries to focus on the evaluation of promising states by refining around these states.

Description

The main motivation of our ISDP approach was the inability of solving higher-dimensional problems with a normal SDP approach. However, we must consider lower dimensional problems to be able to compare both approaches. Hence, we do not allow investments of competitors for this comparison.

Besides the fact that the normal SDP approach considers the whole state space, the two approaches differ slightly in the selection of the fuel price scenarios. In

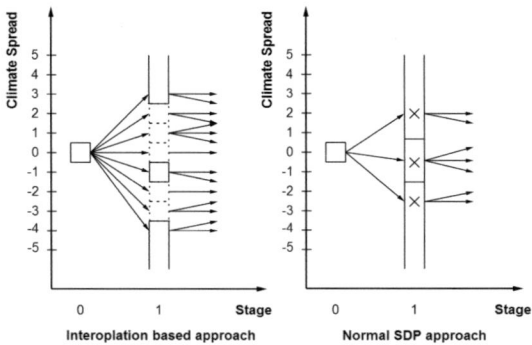

Figure 3.21: Climate spread intervals for the ISDP approach and the normal SDP approach. Dashed boxes indicate non-saved states (ISDP approach), the 'x' marks the selected representative climate spreads (normal SDP approach).

principle, we use the approach described in Subsection 3.2.3 to select the fuel price scenarios for the SDP model. The only difference lies in the interval that a fuel price scenario represents.

In our model, a fuel price scenario covers all fuel prices in a fixed interval around the chosen price. For example, a fuel price scenario of 10.00 incorporates all fuel prices between 9.50 and 10.49. We use intervals of equal size for all fuel price scenarios (except the extreme ones).

For the SDP approach, we do not use fixed intervals. The interval that a fuel price scenario covers depends on the selection of the representative fuel price scenarios. We use Algorithm 1 (page 132) to select a fixed number of representative fuel price scenarios for the first year of each stage. The bounds of each fuel price interval are the middle between the representative fuel price scenario and its two neighbors.

Figure 3.21 illustrates this approach based on the climate spread. The left figure shows the states used in our ISDP algorithm. All climate spread intervals are of size 1, except the two intervals representing the maximum and minimum climate spread.

The right part of Figure 3.21 shows the climate spread intervals of the normal SDP approach. We assume that three representative climate spreads have been selected. These representative climate spreads are -2.5, -0.5 and 2.0. Consequently, the climate spread interval covered by scenario -0.5 is between -1.5 and 0.75.

Besides the interval of each fuel price scenario, there is no difference to our approach in the determination of the contribution margin for a state or in the determination of the transition probabilities. As the normal SDP approach considers

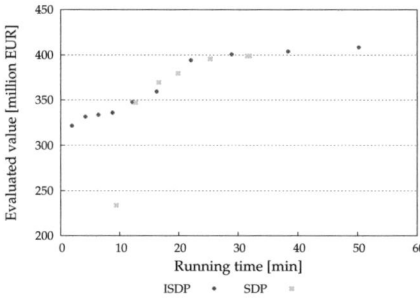 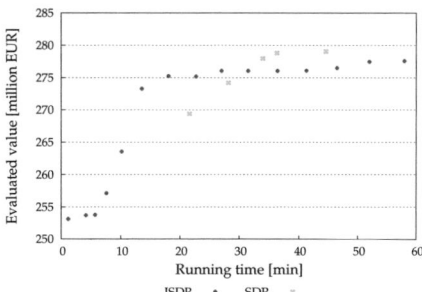

(a) Fuel prices simulated as correlated GBM. (b) Logs of fuel prices simulated as correlated mean-reversion processes.

Figure 3.22: Comparison of the ISDP and the normal SDP approach. The depicted SDP values differ in the number of considered climate spread scenarios.

the whole state space for the solution of the problem, no refinements are required. Consequently, the problem needs to be solved only once.

Comparison

Comparing the ISDP model to a normal SDP approach for models that can be solved with both approaches yields ambiguous results. The advantage of one approach compared to another one depends on the considered problem. As long as the considered problem is solvable with a normal SDP approach, the solutions of both models are of similar quality.

Figure 3.22 visualizes the solution quality and the running time for the ISDP approach and the normal SDP approach. The solutions depicted for the ISDP approach correspond to the solutions obtained after each iteration of the algorithm. The SDP solutions are independent solutions based on a different number of considered climate spread scenarios. The better the solution, the more climate spread scenarios have been considered.

The results in Figure 3.22(a) and 3.22(b) are based on different stochastic processes and lead to different expected profits. However, the performance of the ISDP approach compared to the SDP approach is similar in both cases. The ISDP approach needs less time than the SDP model to find a solution. The quality of the first solution of the SDP (that is based on five different climate spread scenarios per stage) is worse than the solutions of the ISDP approach found during the same time. Increasing the number of considered fuel price scenarios to seven or more, solution quality and running times are comparable for both approaches.

However, the ISDP approach has an advantage, since we do not know the optimal solution of any approach.[19] With our approach, we receive at least a rough idea about the quality of the current solution by comparing it to the quality of previous solutions.

If we apply the normal SDP approach, we need to solve the model from scratch to get another solution based on a different number of fuel price scenarios applicable for comparison.

Consider for example the first solution of the SDP approach in Figure 3.22(a). This solution is based on five climate spread scenarios per stage. The next solution, based on seven climate spread scenarios per stage, is significantly better and requires only slightly more running time. Unfortunately, we have to solve the model from scratch to calculate this additional solution for comparison purposes. We might reuse some of the contribution margins calculated by the fundamental electricity market model, but the SDP problem must be solved again as the number of states and the transition matrix change. Hence, the determination of these two solutions takes somewhere between 13 minutes (the time to solve the model based on seven fuel price scenarios) and 22 minutes (the time to solve both models). With our algorithm, we have already calculated five different solutions after 13 minutes.

The ISDP model has another important advantage compared to the normal SDP approach: the explicitly considered state space is significantly smaller. This yields two benefits.

First, less states have to be evaluated. In our test instances based on the LDC electricity market model, the savings due to the evaluation of less states did not clearly outweigh the overhead of the ISDP model. However, if a computational intensive electricity market model—like the LP models for example—is used, the determination of the contribution margins for the states consumes most computation time. In such a case, the ISDP model is significantly faster.

Second, the ISDP model scales better. As less states are explicitly saved, it requires less memory. Consequently, it can still be used to solve problems that cannot be solved by a normal SDP approach.

---

[19]The solution of the normal SDP approach is optimal for the fuel price scenarios used in the model. As these fuel price scenarios are just a simplification of the real stochasticity of fuel prices, the solution of the normal SDP approach is not optimal in reality.

3.7 Summary

In this chapter, we presented and evaluated an ISDP formulation for the power generation expansion problem. To keep the problem computationally tractable, we introduced several simplifications. The first one is temporal aggregation. To be able to consider construction times of new units without increasing the complexity of the problem, we allow new decisions only every four years. As the focus of our model lies on the strategic level, we believe that this simplification is acceptable.

The second simplification is an aggregation of the coal price, the gas price and the carbon price into one single value, the climate spread. The idea behind this aggregation is that for the unit commitment decisions between CCGT units and coal units, the exact fuel prices are less important than the cost difference between these units. This cost difference is captured by the climate spread. Using the climate spread instead of the exact fuel prices allows us to decrease the dimensionality of the problem by two dimensions. To the best of the author's knowledge, such an aggregation has not yet been used in literature. In our tests, the quality of the solutions obtained with this approach is only slightly worse than the quality of solutions based on an explicit representation of fuel prices. In the same time, the approach based on climate spreads is significantly faster, and allows for a more detailed representation of other aspects of the problem, e.g., external investments.

In contrast to many other approaches in the literature (e.g., Jensen and Meibom [JM08], Klein et al. [Kle+08], Swider and Weber [SW07]), we do not take the perspective of a single, centralized decision maker in the market. Consequently, we have to model the decisions of competitors. We presented two simple approaches, how such investments of competitors can be modeled. The first one is solely based on the scarcity of supply. The second one additionally considers the expected profitability of new units. As the decisions of the competitors depend on the decisions of the considered operator, we need to add extra attributes to the states. These attributes are used to represent the number of new units built by competitors. Compared to other models that ignore these decisions, the consideration of external investments makes our model more complex.

To be able to solve this complex problem, we use an adaptive grid method. The adaptive grid method is the main difference of our model compared to existing approaches in literature. Instead of solving the original problem, we solve an approximated problem considering only a subset of the state space. To improve the solution quality, we adaptively refine the grid.

These adaptive refinements are based on the splitting priority assigned to each hyperrectangle of the grid. For the calculation of this splitting priority, we presented and compared different splitting criteria from literature with a new splitting criterion. It turns out that for our problem, a combination of a value function based splitting criterion and our PIOS criterion works best on average.

While we iteratively solve approximated problems, we consider only a subset of the state space, but the whole decision space. Hence, we have to approximate the value of states that are not explicitly considered. For this purpose, we tested two different interpolation methods: multilinear interpolation and simplex-based interpolation. For our problem, the more accurate results of multilinear interpolation outweigh the faster computation times of simplex-based interpolation.

Finally, we compared the performance of our algorithm with two alternative approaches, the deterministic approach based on expected values for the stochastic fuel prices, and a normal SDP approach. Our algorithm clearly outperforms the deterministic approach. Compared to the deterministic approach, more of the less-capital intensive CCGT units are built.

The solutions of the normal SDP problem and our algorithm are similar concerning the quality of the solutions as well as the running time. However, our algorithm has two important advantages, that cannot be captured by our tests. First, our algorithm scales better than the normal SDP approach as it does not consider all states of the state space. For the test data, we had to select the problems such that they could still be solved with the normal SDP approach. To be able to solve more realistic problems, e.g., considering investments of competitors, we need to use the ISDP approach.

The second advantage of our algorithm is also related to the consideration of a subset of the state space. With our algorithm, we only have to calculate the contribution margins for the considered subset of the state space. With the LDC-based electricity market model, which we used for the comparison, the calculation of the contribution margins for a state is relatively fast. Consequently, the advantages of evaluating only a subset of the states is modest. However, if we replace the LDC model with a (significantly) slower LP model, our algorithm will also be much faster for small problems than the normal SDP approach. For example, the case study evaluating the profitability of CCS units, which we present in Section 5.5, would take much longer if a normal SDP approach is used (if it could be solved at all).

While the ISDP formulation presented in this chapter allows us to solve the power generation expansion problem considering many different fuel price scenarios, the ISDP model has its limitations. First, we have to apply temporal aggregation to consider construction lead times of new plants. Consequently, we cannot model annual decisions. Second, the ISDP approach does not break the curse of dimensionality of DP. It just allows us to push the limits a bit further, but we are still restricted to low dimensional problems. Hence, we cannot consider many different types of new units in our problem. In the next chapter, we describe an investment model based on an alternative method that might circumvent these problems: approximate dynamic programming (ADP).

Chapter 4

Investment Model Based on Approximate Dynamic Programming

In this chapter, we describe an investment model based on ADP, which allows us to consider annual investments while taking into account fuel and carbon price uncertainties. ADP is a technique that gained in importance during the last years. It is best described as a combination of simulation, optimization and statistics. ADP has been successfully applied for example in the area of transportation and logistics (Powell et al. [Pow+02], Simão et al. [Sim+08]), finance (Tsitsiklis and Van Roy [TV01]), inventory management (Van Roy et al. [Van+97]), for discrete routing and scheduling problems (Spivey and Powell [SP04]) and for batch service problems (Papadaki and Powell [PP03]).

The main idea of ADP is to make decisions based on an approximation of the value function instead of the exact value function, as it is done in DP. It is assumed that the value of a state can be described as a function of some of its attributes. For example, the value of a state might depend on the number of power plants. A wide variety of different functions can be used to approximate the value of a state. In our model, we use a relatively simple, but popular choice: piecewise-linear function approximations.

Making decisions based on approximated values has two advantages, which are discussed in detail later on in this chapter: First, it avoids the complete enumeration of the state space. Second, it allows us to choose the value function approximation such that it can be stored more efficiently than the lookup table used in normal DP. As a consequence, ADP can tackle high-dimensional problems that cannot be solved with DP.

In this chapter, we examine how ADP can be applied to solve the power generation expansion problem. While the idea is appealing, we encounter several problems. First, the choice of a state representation and the modeling of the information flow is more complicated than in normal SDP. Second, there are many design choices that need to be tuned for the specific problem. A wrong choice can result in a diverging algorithm or in an algorithm that converges to a suboptimal solution. But even if we find the right values for these parameters, we have no guarantee that the algorithm converges to the optimal solution. To be able to assess the quality of the solutions, we compare the solutions of the ADP approach with the solutions of our ISDP model. Due to its ability to consider annual investments, the solutions of the ADP model are slightly better than those of the ISDP model.

The remainder of this chapter is organized as follows: We start with an introduction to ADP. Thereby we present the most important ideas of ADP, and point out to the links and differences to DP. Next, we show how ADP can be applied to solve the power generation expansion problem. Then, we discuss different problems of the ADP formulation of our investment model. Finally, we evaluate the solutions of our ADP formulation and conclude the chapter with a summary.

4.1   Approximate Dynamic Programming

ADP deals with large-scale stochastic problems evolving over time. Such problems arise in many disciplines, including operations research, artificial intelligence and engineering. Different communities tackled such problems within their own context. As a result, similar solution approaches with different names and notations have been developed. These names are *reinforcement learning*, *neuro-dynamic programming* and *approximate dynamic programming*.

In the artificial intelligence community, the methods for solving Markovian Decision Problems became known as *reinforcement learning*. Thereby, an agent learns through trial-and-error interactions with a dynamic environment. The agent selects an action that changes the state of the environment. Based on the new state of the environment, the agent receives a reward or a penalty. Through this "reinforcement" signal, the agent learns whether its decision was good or not. The objective of the agent is to maximize the reward over time.

Reinforcement learning is mainly applied to robot control, but also for game playing (e.g., Backgammon, Go, Chess). An introduction to reinforcement learning has been written by Sutton and Barto [SB98], for an extensive study covering the most important aspects of reinforcement learning see Kaelbling et al. [Kae+96].

In the control-theory community, the term *neuro-dynamic programming* is used for a solution method of control and sequential decision making under uncertainty. The name neuro-dynamic programming has been chosen to reflect the reliance on concepts of dynamic programming and neural networks. As ADP, neuro-dynamic programming approximates the value-function of a state. This approximation is based on a neural network approach. An introductory textbook to neuro-dynamic programming has been written by Bertsekas and Tsitsiklis [BT96].

The term *approximate dynamic programming* is mainly used in the operations research community. Similar to neuro-dynamic programming, it stresses the connection to dynamic programming. However, the term approximate is more general: the approximation of the value function may be based on aggregation, on linear or piecewise-linear functions, or also on neural nets.[1] A detailed description of ADP can be found in the textbook by Powell [Pow07].

In the following, we give a short introduction to ADP based on the book by Powell. We start with a description of the main ideas of ADP: forward dynamic programming and value function approximation. Next, we outline a basic ADP algorithm, before we discuss several possibilities for the value function approximation. Finally, we present different algorithms that can be used to solve ADP problems.

4.1.1 Basic Idea

Dynamic programming is suited to solve low-dimensional dynamic problems efficiently. However, the state space of a problem grows often exponentially with the number of dimensions. This problem is known as the curse of dimensionality, and is a significant obstacle for the application of DP to high-dimensional problems.

Powell points out that there are actually three different curses of dimensionality. These are:

- The state space: If a state variable has I dimensions, and can take on L possible values per dimension, we have up to L^I different states.

---

[1]The terms neuro-dynamic programming and approximate dynamic programming are sometimes used synonymously. In their overview of neuro-dynamic programming, Bertsekas and Tsitsiklis [BT95] use the term "neural network" as a synonym to "approximating architecture".

- The outcome space of the stochastic parameters: The random variable may have J dimensions with M possible outcomes. Then, there might be up to M^J outcomes.

- The action space: The decision vector might have K dimensions with N outcomes. Then, there are up to N^K outcomes.

Applying the backward induction approach to solve an SDP problem requires an iteration over the whole state space. For each state, we solve the equation:

$$
\begin{aligned}
V_t(x_t) &= \max_{a_t \in \Gamma(x_t)} \left(F(x_t, a_t) + \beta \cdot \mathbb{E} \left\{ V_{t+1}(x_{t+1}) \mid x_t \right\} \right) \\
&= \max_{a_t \in \Gamma(x_t)} \left(F(x_t, a_t) + \beta \cdot \sum_{x_{t+1} \in \mathcal{X}_{t+1}} P(x_{t+1} \mid x_t, a_t) \cdot V_{t+1}(x_{t+1}) \right).
\end{aligned}
\tag{4.1}
$$

The solution of problem (4.1) requires an iteration over all possible decisions and all outcomes of the stochastic parameter. For many real-world problems, such an enumeration is computationally intractable.

4.1.1.1 Forward Dynamic Programming

The main idea of ADP is to replace the exact value function $V_{t+1}(x_{t+1})$ with an approximation, to which we refer as $\tilde{V}_{t+1}(x_{t+1})$. Instead of computing the optimal solution of equation (4.1) stepping from the last stage to the first one, ADP goes forward in time, and makes decisions based on value function approximations. ADP starts with the initial state, and makes a decision based on the solution of the following problem:

$$
a_t^*(x_t) = \operatorname*{argmax}_{a_t \in \Gamma(x_t)} \left(F(x_t, a_t) + \beta \cdot \mathbb{E} \left\{ \tilde{V}_{t+1}(x_{t+1}) \mid x_t \right\} \right).
\tag{4.2}
$$

The optimal decision of this equation, and a sample realization of the random variables determine the next state. The new information that is gathered through solving problem (4.2) and sampling the random variables is used to update the value function approximations. To obtain a good approximation of the value function, this approach is repeated iteratively.

During each iteration, the current value function approximation is updated with the new information. Assume that we are in state x_t in iteration n, and the value function approximation of x_t based on the information available up to the last iteration is $\tilde{V}^{n-1}(x_t)$. Let $\hat{v}_t^n$ be the expected value of the optimal decision for state

x_t (based on equation (4.2)). Then, we update the value function approximation as follows:

$$\tilde{V}^n(x_t) = (1 - \alpha_{n-1}) \cdot \tilde{V}^{n-1}(x_t) + \alpha_{n-1} \cdot \hat{v}_t^n.$$

Thereby, the parameter α is the stepsize, which we are going to discuss later on in this section.

While forward dynamic programming avoids the complete enumeration of the state space, there are several difficulties. The first problem is to get a good value function approximation of the relevant states. Assuming that we use a lookup table representation, i.e. we save the value of each state in a lookup table, we only update the value function approximation of states we visit. However, we might never visit some interesting states, as the value function approximation is too low. We deal with this problem in Subsection 4.1.3, where we describe alternative ways of approximating the value function of a state.

The second problem with formulation (4.2) is that it still requires the calculation of an expectation. There are two issues with the calculation of this expectation: First, we have to save the transition probabilities, and second, we have to iterate over all succeeding states to calculate the expected value. Following Powell [Pow09], this is the most problematic curse of dimensionality. Powell circumvents this problem by introducing the post-decision variable.

4.1.1.2 Post-Decision Variable

Instead of using one state variable, we might use two: one variable directly before a decision is taken, and one variable indicating the state directly after the decision is taken, but before any new information about the realization of the stochastic parameters is available. We denote the first variable, the pre-decision variable, with x_t, while the second one, the post-decision variable, is denoted with x_t^a. Figure 4.1 illustrates this approach.

The introduction of the post-decision variable allows us to break the transition function $x_{t+1} = T(x_t, a_t, w_{t+1})$ into two parts with $x_t^a = T^a(x_t, a_t)$ and $x_{t+1} = T^w(x_t^a, w_{t+1})$.

The advantage of this approach is that it allows us to split problem (4.2) in two parts: a deterministic optimization problem, and a simple simulation. The optimization problem is the following:

$$a_t^*(x_t) = \underset{a_t \in \Gamma(x_t)}{\operatorname{argmax}} \left(F(x_t, a_t) + \beta \cdot \tilde{V}_t(T^a(x_t, a_t)) \right). \tag{4.3}$$

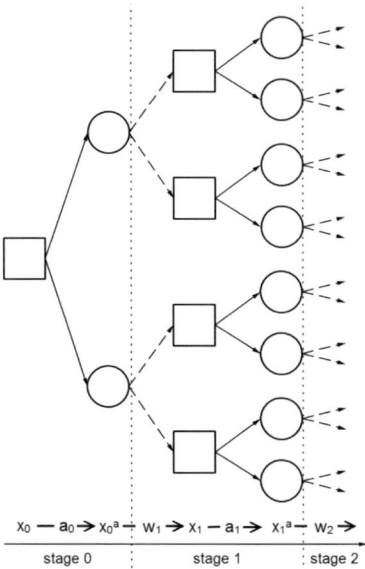

Figure 4.1: A decision tree using post-decision states. Pre-decision states are shown as squares, post-decision states as circles. Solid lines are decisions, dotted lines are the outcomes of random events.

Note that the value function approximation is now based on the post-decision variable. The main difference between problems (4.2) and (4.3) is that we no longer have to calculate the expectation in the optimization problem. After we solve problem (4.3), we sample an outcome of the random variable, and determine the next pre-decision state $x_{t+1} = T^w(x_t^a, w_{t+1})$.

4.1.2 Basic Algorithm

In this section, we are going to discuss several design decisions of ADP based on a generic ADP algorithm (Algorithm 5). This algorithm is taken from Powell [Pow07, p. 106].

The algorithm starts with an initialization of the value function approximation. Then, the main part of the algorithm is run for a fixed number of iterations N. We start with choosing a sample path of the random variables. In our problem, this sample path is one realization of the stochastic fuel and carbon prices for the whole optimization horizon. Then, we determine the optimal decision (line 5) of the current pre-decision state. The choice of this optimal decision is based on

---

Algorithm 5 Generic ADP algorithm using the post-decision variable

---

1: Initialize $\tilde{V}_t^0(x_t)$ for all states x_t.
2: **for** $n = 1$ **to** N **do**
3:     Choose a sample path w^n.
4:     **for** $t = 0$ **to** T **do**
5:         $a_t^n = \underset{a_t \in \Gamma(x_t^n)}{\operatorname{argmax}} \left(F(x_t, a_t) + \beta \cdot \tilde{V}_t \left(T^a(x_t, a_t) \right) \right)$
6:         $\hat{v}_t^n = F(x_t, a_t^n) + \beta \cdot \tilde{V}_t(T^a(x_t, a_t^n))$
7:         **if** $t > 0$ **then**
8:             $\tilde{V}_{t-1}^n(x_{t-1}^{a,n}) = (1 - \alpha_{n-1}) \cdot \tilde{V}_{t-1}^{n-1}(x_{t-1}^{a,n}) + \alpha_{n-1} \cdot \hat{v}_t^n.$
9:         **end if**
10:       $x_t^{a,n} = T^a(x_t^n, a_t^n)$
11:       $x_{t+1}^n = T^w(x_t^{a,n}, w_{t+1}^n)$
12:     **end for**
13: **end for**
14: **return** $\tilde{V}^N$

---

the contribution function and the value function approximation of the succeeding post-decision state.

The contribution function depends on the current (pre-decision) state and the decision we choose. In our problem, this contribution function is composed of the contribution margins for the electricity production of new power plants less the investment costs for new plants. The value function approximation of the post-decision state is an expectation of the value of being in this state. It depends on the succeeding decisions, as well as on the outcomes of the stochastic variables.

Next, we determine the expected value of the optimal decision (line 6). This value is used to update the value function approximation of the preceding post-decision state (line 8). Then, we select the next post-decision state based on the chosen decision (line 10), and the next pre-decision state, which depends on the post-decision state and the outcome of the stochastic variable (line 11).

Algorithm 5 is not very specific in some points, which we are going to address in the following. These points are the initialization of the value function approximations and the choice of the stepsizes. Furthermore, we are going to discuss different possibilities, how it can be avoided that the algorithm explores only a small subset of the state space. This problem is known as the exploration vs. exploitation problem.

4.1.2.1 Initialization

Before we start with the main part of the algorithm, we have to determine initial value function approximations. Our objective is to start with value function approximations that are close to the real value functions. However, as we have only limited knowledge about the problem, we can only guess the real value functions in most problems.

There are basically two different approaches to initialize value function approximations: We might either underestimate the values or overestimate them. Both strategies result in a different behavior of the algorithm.

Assume that we underestimate the values. In our problem, we might start with initial value function approximations of zero for each state (which will be in most cases an underestimation). In that case, the decision we chose depends only on the contribution function (line 5 of Algorithm 5). Consequently, the optimal decision will always be to build no new units, as this is the only decision that does not involve any costs. No update of the value function approxmation takes place, as the value function for the optimal decision is correct. Obviously, if the optimal solution of the problem involves the construction of new units, we will not find it with Algorithm 5 and an initialization of zero for the value function approximations.

An overestimation of the value of a state results in an opposite behavior. The expected future value of a state dominates the value of the contribution function. The solution of the first iteration will be the same as with an initialization of zero (assuming that all value functions are initialized with the same value). After the first iteration, the value function approximations of the visited states are decreased, and in the next iteration, a new solution is chosen.

If we consider a deterministic problem, this approach always leads to the optimal solution. However, optimality is not guaranteed in a stochastic setting. The disadvantage of overestimating the value functions is that it results in the exploration of many states. We might even iterate several times over the whole state space, before the correct values are found. Nevertheless, Algorithm 5 combined with an initialization that overestimates the value function approximations is a popular algorithm in the artificial intelligence community, known as A* algorithm (Hart et al. [Har+68]).

4.1.2.2 Exploration vs. Exploitation

The two different initialization strategies can be seen in a more general context as the problem between exploration and exploitation. This is a common problem that also occurs in many other algorithms like meta-heuristics. Thereby, exploration is used to gather new information, even if the actions required to do so are not optimal. If the chosen decisions are optimal based on the current knowledge, this knowledge is "exploited", hence one refers to this as exploitation.

Algorithm 5 is written as a pure exploitation algorithm. In line 5, we always select the best decision based on the current value function approximations. We use this decision in line 10 to determine the next state. As we have illustrated with the simple example of an underestimation of initial value functions, we might never find the optimal solution, because we never visit underestimated states.

A pure explorative strategy would replace the optimal decision in line 10 with an arbitrary decision. However, the drawback of this approach is that it corresponds to a pure random search, which is not efficient. In general, it is a very important, but also a very difficult task to find a good balance between exploration and exploitation.

A simple strategy to mix exploration and exploitation consists in defining an exploration rate ρ specifying the ratio of decisions that are chosen randomly. More sophisticated strategies are the following two strategies:

- **ϵ-greedy exploration** uses instead of a fixed exploration rate a variable exploration rate depending on the current iteration. With an increasing number of iterations, the tendency for exploration is decreased. The idea behind this decreasing probability for exploration is the assumption that with an increasing number of iterations, the value function approximations get more accurate. Consequently, the need for exploration is decreased.

  The probability for an exploration step at iteration n is denoted with

  $$P^n(x_t) = \epsilon^n(x_t).$$

  Powell suggests to use $\epsilon^n(x_t) = c/N^n(x_t)$ where $0 < c < 1$, and $N^n(x_t)$ is the number of times state x_t has been visited up to iteration n. If an exploration step is taken, each decision is chosen with the same probability.

- **Boltzmann exploration** uses the expected value of a decision to assign a probability to each possible decision. Following the same reasoning as ϵ-

greedy exploration, the probability of exploration steps is decreased with time.

Boltzmann exploration uses the following formula to calculate the probability for any possible action:

$$P(a_t \mid x_t) = \frac{e^{Q(x_t, a_t)/T}}{\sum\limits_{a'_t \in \Gamma(x_t)} e^{Q(x_t, a'_t)/T}}.$$

Thereby, $Q(x_t, a_t)$ corresponds to the expected value of decision a_t, which is $Q(x_t, a_t) = F(x_t, a_t) + \tilde{V}(x_t^a)$. The parameter T is known as temperature. In general, a high value of T is chosen during initial iterations, and with increasing number of iterations, T is steadily decreased to account for more accurate value function approximations. Powell proposes the following formula to calculate the temperature depending on the current state x_t and iteration n:

$$T^n(x_t) = \frac{\max\limits_{a_t \in \Gamma(x_t)} |Q(x_t, a_t^*) - Q(x_t, a_t)|}{N^n(x_t)}, \tag{4.4}$$

where a_t^* denotes the optimal decision at state x_t. Applying formula (4.4), the temperature is equal to the difference in the expected value of the best and the worst decision, divided by the number of times the current state has been visited.

4.1.2.3 Stepsizes

A very important parameter of an ADP algorithm is the choice of the stepsize rule. Powell [Pow07, p. 184] states that "it is possible to find problems where provably convergent algorithms simply do not work, and the only reason is a poor choice of stepsizes." The choice of the stepsize rule has also an important influence on the convergence speed of the algorithm.

An extensive comparison of different stepsize rules is given by George and Powell [GP06]. In the following, we describe some of the most popular stepsize rules. Thereby, n denotes the current iteration of the algorithm.

- **Constant stepsizes** are the simplest kind of stepsize rule. The stepsize is defined as

$$\alpha_{n-1} = \begin{cases} 1 & \text{if } n = 1, \\ \overline{\alpha} & \text{otherwise,} \end{cases}$$

with $\overline{\alpha}$ being a constant value. In general, declining stepsizes are preferred to constant ones in ADP. As the accuracy of the value function approximation generally increases with time, declining stepsizes are more appropriate than constant ones.

- **Generalized harmonic stepsizes** belong to the class of declining stepsizes and are defined as:

$$\alpha_{n-1} = \frac{a}{a + n - 1}.$$

The parameter a determines how fast the stepsize declines. Increasing a leads to a slower decrease of the stepsize.

- **Polynomial learning rates** are similar to generalized harmonic stepsizes. They are defined as:

$$\alpha_{n-1} = \frac{1}{n^\beta},$$

with $\beta \in (0.5, 1]$. Smaller values of β result in a slower decline of the stepsize.

- The **Search-then-converge stepsize rule** (Darken and Moody [DM90]) is a very versatile stepsize rule. Powell uses the following definition:[2]

$$\alpha_{n-1} = \alpha_0 \cdot \frac{\frac{b}{n} + a}{\frac{b}{n} + a + n^\beta}.$$

The parameters a, b and β can be used to adjust the speed with which the stepsize declines. The stepsize rule is named "search-then-converge", as it allows an initial period with high stepsizes (the search period), before the stepsizes are reduced to achieve convergence.

Unfortunately, there is no single best stepsize rule. The performance of stepsize rules is problem dependent. It depends furthermore on the choice of the different parameters of the stepsize rules, which also depend on the considered problem.

---

[2]This definition is slightly different from the original proposal.

4.1.3   Value Function Approximations

So far, we have not specified how the value function approximations are saved. We just assumed that there is a value function approximation for each state, which can be used and updated.

The standard way of saving values of states in DP is a lookup table representation. For every state in the state space, the expected value is explicitly saved. While this approach might be suitable for small problems, it is intractable for many real world problems where the state space gets too large.

In this subsection, we present three different methods of approximating the value function without explicitly saving the value of each state. The methods we present are based on the idea that the value of a state can be described as a function of a few attributes of the state.

We start with a description of the simplest case: a linear relationship between the number of attributes and the value of a state. However, for many problems such a linear relationship is too simplistic. A more realistic approach for many problems is a piecewise-linear value function approximation. We present this approach in the second part of this subsection.

There are many more possible approximation architectures that might be used, for example neural nets (see Bertsekas and Tsitsiklis [BT96]). We restrict our small introduction to the before mentioned methods, as the piecewise-linear value function approximation is the basis for the method we apply, and linear value function approximations are a simple special case of piecewise-linear value function approximations.

After presenting these two approaches, we briefly describe how these approaches can be combined with the standard lookup table approach, to be able to consider the influence of attributes that cannot be described with a function approximation. This indexed piecewise-linear value function approximation is the method we choose for our problem. In the last part of this subsection, we discuss advantages and difficulties of value function approximations.

4.1.3.1   Linear Approximations

A linear value function approximation assumes that there is a linear or nearly linear relationship between the attributes of a state and the value of this state. For example, if we consider a resource allocation problem, one might assume that the value of an additional resource is constant, independent of the current number of resources we have.

If such a linear relationship exists, we do not have to explicitly save the value of each state. Instead, we can use a linear value function approximation. Let R_{ti} describe the number of resources of type i at time t, then we can use the following function to approximate the value:

$$\tilde{V}_t(x_t) = \sum_{i \in \mathcal{I}} \tilde{v}_{ti} \cdot R_{ti},$$

where $\mathcal{I}$ is the set of different resources (e.g., machines), and $\tilde{v}_{ti}$ is the approximated value of an additional resource of type i.

Assume that we have I different resources, where each of the resources can take up to L different values. Using a lookup table, we would have to save the values of L^I states. Using a linear value function approximation, we only have to save I values (the value of an additional resource of each type).

If we use a linear value function approximation instead of a lookup table, the update mechanism of our algorithm slightly changes. Instead of updating the value of being in a specific state, we update the slope of the value function approximation. This can be done by defining $\hat{v}_t$ as the gradient of the value function with respect to R_t:

$$\hat{v}_t = \begin{pmatrix} \hat{v}_{ti_1} \\ \hat{v}_{ti_2} \\ \vdots \\ \hat{v}_{ti_{|\mathcal{I}|}} \end{pmatrix},$$

where

$$\hat{v}_{ti_j} = \frac{\partial V_t(R_t)}{\partial R_{ti_j}}.$$

To obtain the elements of the gradient, we can use numerical derivatives. Once the gradient is computed, $\tilde{v}_{t-1}$ can be updated following

$$\tilde{v}_{t-1}^n = (1 - \alpha) \cdot \tilde{v}_{t-1}^{n-1} + \alpha \cdot \hat{v}_t.$$

4.1.3.2 Piecewise-Linear Approximations

For most real world problems, a linear relationship between the number of resources and the value of a state is too simplistic. Instead of remaining constant, the value of an additional resource is often diminishing. In economics, this concept in known as the *law of diminishing marginal returns*.

In such a case, a piecewise-linear value function approximation can be used instead of a linear one. We can write the value function approximation as

$$\tilde{V}_t(x_t) = \sum_{i \in \mathcal{I}} \tilde{V}_{ti}(R_{ti}).$$

When a linear value function approximation is used, only one slope per attribute has to be estimated. Now, we have to estimate multiple slopes per attribute, which is expressed by $\tilde{V}_{ti}(R_{ti})$.

Let $\tilde{V}_t(R_t)$ be the slope to the right of R_t. If we assume that the value of an additional resource is diminishing, then the slopes of our value function approximation are monotonically decreasing, i.e., $\tilde{V}_t(R_t) \geq \tilde{V}_t(R_t + 1)$. The standard updating mechanism would be similar to the one used for the linear approximation:

$$\tilde{V}_{t-1}^n(R_t) = (1 - \alpha) \cdot \tilde{V}_{t-1}^{n-1}(R_t) + \alpha \cdot \hat{v}_t(R_t).$$

Note that $\tilde{V}_{t-1}^n$ and $\hat{v}_t$ depend now on R_t, as we no longer use a constant slope per attribute. If we apply this updating mechanism, the resulting slopes of the value function approximation may no longer be monotonically decreasing.

To enforce the monotonicity of the slopes, we apply the separable, projective approximation routine (SPAR) proposed by Powell et al. [Pow+04]. The SPAR algorithm proceeds the following way: Assume that we sample $r = R^n$ and obtain the estimate of the slope $\tilde{V}^n(R^n)$. Then, we determine the normal updates (ignoring possible violations of the monotonicity) and save them in the vector $\tilde{y}^n(r)$ as follows:

$$y^n(r) = \begin{cases} (1 - \alpha_{n-1}) \cdot \tilde{V}^{n-1}(R^n) + \alpha_{n-1} \cdot \hat{v}^n(R^n) & \text{if } r = R^n, \\ \tilde{V}^{n-1}(r) & \text{otherwise.} \end{cases}$$

For all slopes different from r, we use the approximations from the previous iteration. We update only the slope to the right of r.

If an update violates the monotonicity, then either $\tilde{y}^n(R^n) < \tilde{y}^n(R^n + 1)$ or $\tilde{y}^n(R^n - 1) < \tilde{y}^n(R^n)$. We fix this violation by solving the problem:

$$\begin{aligned} \min_{v} &\ \|v - \tilde{y}^n\|^2 \\ \text{s.t.} &\quad v(r+1) - v(r) \leq 0 \qquad \forall r. \end{aligned} \tag{4.5}$$

The solution of problem (4.5), v, is the feasible solution that has the minimal distance to $\tilde{y}^n$.

The solution of this problem is simple. Assume that we have a monotonicity violation to the left of R^n, i.e., $\tilde{y}(R^n) > \tilde{y}(R^n - 1)$. To resolve this violation, we set $v(R^n - 1) = v(R^n) = 1/2 \cdot (\tilde{y}(R^n) + \tilde{y}(R^n - 1))$. If v is a monotone decreasing function now, we are done. Otherwise, we have a violation to the left of $R^n - 1$. Then, we continue with averaging until monotonicity is restored.

If the monotonicity violation is to the right of R^n, we use the same approach averaging the values to the right of R^n. The solution of problem (4.5) is our new value function approximation $\tilde{V}^n$.

4.1.3.3 Indexed (Piecewise) Linear Approximations

There are problems where the value of some, but not of all attributes can be described by a linear or a piecewise-linear function approximation. For such problems, the linear or piecewise-linear function approximation approach can be combined with the lookup-table approach. This method is called the indexed (piecewise) linear value function approximation approach.

Assume that a state consists of several attributes, e.g., some resources and some other attributes that represent information about the price of raw materials. For the attributes representing the different resources, we might find a linear or piecewise-linear relationship such that we can use a function approximation. However, this function approximation may depend on the price of raw materials. In such a case, we can use separate function approximations for each price scenario.

Let $\phi(x_t)$ be the value of the attributes that cannot be expressed as a function approximation. We assume that these values are discretized into a set Φ_t, which should not get too large. Then, instead of using one value function approximation for each of the attributes that can be expressed as a linear or piecewise-linear function, we use for every of these attributes a separate value function approximation for each element of Φ_t. The value function approximation is then (using a piecewise-linear function approximation)

$$\tilde{V}_t(x_t) = \sum_{i \in \mathcal{I}} \tilde{V}_{ti}(R_{ti} \mid \phi(x_t)).$$

4.1.3.4 Advantages and Problems

An important advantage of value function approximations is the compact representation, which requires significantly less memory than a lookup table representation. However, there is a second advantage that might not be so obvious.

Using a lookup table representation, an update of the value function concerns only the currently visited state. As a consequence, we obtain only value function approximations for states we have visited. Hence, we need to iterate over the whole state space to get a value function approximation for all states.

If we use a linear or a piecewise-linear value function approximation, an update of the slope of the value function results in a better approximation for many states. Thus, we do not need to iterate over the whole state space to solve the problem.

Besides the requirement of a linear or a piecewise-linear value function, there is a second condition, which must be fulfilled if a linear or piecewise-linear value function approximation should be applied: the attributes used for the approximation must be independent. Otherwise, the approach does not work, as we calculate the value of a state as the sum of the value of the different attributes. If a state consists of several attributes, where some of them are not independent, we might use the indexed (piecewise) linear value function approximation approach.

4.1.4 Algorithms

There are many different algorithms that can be used to solve ADP problems. Many of these algorithms are variations of algorithms that are commonly applied to solve Markov decision problems. Examples of such algorithms are the value iteration and the policy iteration.

In this subsection, we first briefly describe two possible choices of ADP algorithms concerning the timing of the value function approximation updates. Next, we describe the value iteration and the policy iteration. Then, we present a combination of the value iteration and the policy iteration, the Monte Carlo value policy iteration. This is the algorithm we use to solve the power generation expansion problem. A description of many other algorithms can be found for example in [Pow07].

4.1.4.1 Single-Pass vs. Double-Pass

One design decision of an ADP algorithm is the timing of the value function updates. There are basically two possibilities: either the update takes place immediately after the new estimation of the value of a state is available, or the update takes place after each iteration.

The first approach corresponds to the one we have seen in Algorithm 5 and is know as *single-pass* procedure. Applying this approach, the value function approximation of the previous state is updated before the algorithm proceeds to the next stage. The single-pass approach is simple to implement, but not necessarily very efficient.

The alternative method, the *double-pass* procedure, is in general more efficient. Instead of performing updates immediately, the algorithm proceeds to the last stage without any updates. Then, the algorithm starts with the value function updates at the last stage, stepping back until the initial state is reached. The advantage of this approach is that the information gathered at later stages is already used for the update of states at earlier stages.

4.1.4.2 Value Iteration

Value iteration is one of the most popular algorithms used to solve Markov decisions problems. It is similar to the backward induction approach which we use to solve our ISDP formulation presented in the previous chapter. The difference is that value iteration can also be used to solve infinite horizon problems.

The basic ADP algorithm (Algorithm 5 on page 189) is a variant of the value iteration. During each iteration, we estimate and update the value functions. The difference between this ADP value iteration and the standard value iteration used for Markov decision problems is the following: The standard value iteration updates the value functions for all states during each iteration, the ADP version only those of some states.

The ADP version selects one sample path of the stochastic parameters per iteration. It starts at the initial state, takes a decision and—based on this decision and the realization of the stochastic process—moves to the next state. This procedure is repeated until a state of the last stage is reached. Only the value functions of visited states are updated. Once a state at the last stage is reached, we go back to the initial state, and repeat these steps with another sample path of the stochastic parameters. The value iteration is run until the value functions converge.

4.1.4.3 Policy Iteration

A disadvantage of the value iteration is that it can take many iterations until the value functions converge and the value iteration stops. Often, the optimal policy converges faster than the value functions. The policy iteration, another popular method to solve Markov decision problems, is based on this idea.

The policy iteration consists of two steps: policy evaluation and policy improvement. We start with an initial policy π_0, and determine the value of this policy. For this purpose, we fix the policy and calculate the value function for each state as follows:

$$V^{\pi_n}(x) = F(x, \pi_n(x)) + \beta \cdot \sum_{x'} P(x' \mid x, \pi_n(x)) \cdot V^{\pi_n}(x').$$

Thereby, $\pi_n(x)$ denotes the decision that is followed by the current policy in state x. The calculated value functions are then used to update the policy as follows:

$$\pi_{n+1}(x) = \operatorname*{argmax}_{a \in \Gamma(x)} \left(F(x, a) + \beta \cdot \sum_{x'} P(x' \mid x, a) \cdot V^{\pi_n}(x') \right).$$

If the new policy corresponds to the previous one, then the algorithm has converged. Otherwise, we have to evaluate the new policy, and update the policy again, until it does not change any more.

As the value iteration, also the policy iteration requires an iteration over the whole state space, and is therefore only applicable for smaller state spaces. An ADP version of the policy iteration can be found in [Pow07, p. 283]. The basic idea is similar to the one used for the ADP version of the value iteration: instead of iterating over the whole state space, we sweep forward in time applying a chosen policy. Only the value functions of the visited states are updated.

4.1.4.4 Monte Carlo Value Policy Iteration

In general, the policy iteration converges faster in terms of iterations than the value iteration. However, the evaluation of the policy takes much longer than the value iteration updates. The *modified policy iteration* algorithm (Puterman and Shin [PS78]) replaces the exact calculation of V^{π_n} with an approximation obtained by performing some modified value iteration steps where the policy is fixed for some iterations. In practice, the modified policy iteration can result in substantial speed-ups.

Powell uses the same idea for the Monte Carlo value and policy iteration algorithm, which is described by Algorithm 6. The outer loop of the algorithm with the index n corresponds to the policy-improvement iterations. The loop with the

---

Algorithm 6 Monte Carlo value policy iteration

1: Initialize $V_t^{\pi,0}$ $\quad \forall t \in T$.
2: Set $n = 1$.
3: **for** $n = 1$ **to** N **do**
4: $\quad$ **for** $m = 1$ **to** M **do**
5: $\qquad$ Choose a sample path w^m.
6: $\qquad$ **for** $t = 1$ **to** T **do**
7: $\qquad\qquad a_t^m = \underset{a_t \in \Gamma(x_t^m)}{\text{argmax}} \left(F(x_t^m, a_t) + \beta \cdot V_t^{\pi,n-1}(T^a(x_t^m, a_t)) \right)$
8: $\qquad\qquad \hat{v}_t^m = F(x_t^m, a_t) + \beta \cdot V_t^{\pi,n-1}(T^a(x_t^m, a_t^m))$
9: $\qquad\qquad x_t^{a,m} = T^a(x_t^m, a_t^m)$
10: $\qquad\qquad x_{t+1}^m = T^w(x_t^{a,m}, w_{t+1}^m)$
11: $\qquad\qquad \tilde{V}_{t-1}^m \leftarrow U^V(\tilde{V}_{t-1}^{m-1}, x_{t-1}^m, \hat{v}_t^m)$
12: $\qquad$ **end for**
13: $\quad$ **end for**
14: $\quad V_t^{\pi,n}(x_t) = \tilde{V}_t^M(x_t)$ $\quad \forall x_t$
15: **end for**
16: **return** $V^{\pi,N}$

---

index m is used to carry out M value iterations to obtain an approximation of the value functions of the current policy.

The main difference between the Monte Carlo value and policy iteration (Algorithm 6) and the value iteration (Algorithm 5) can be found in lines 7 and 8 of Algorithm 6. The value function approximations used to determine a_t^m and $\hat{v}_t^m$ are fixed during M iterations. This corresponds to fixing the policy that is followed, as fixing V^π determines the decision that is taken (depending on the state). After M iterations, the value function approximations V^π are updated with $\tilde{V}$.

Another difference between Algorithm 5 and Algorithm 6 can be found in line 11 of Algorithm 6. However, this difference is only a difference in notation. When we introduced Algorithm 5, we assumed that a lookup table representation is used to save the value function approximations. Hence, the way we defined the value function updates was adopted to this notation. As we have seen, value function updates of linear or piecewise-linear value function approximations are different from the updates of lookup table based value functions. Line 11 of Algorithm 6 refers to the update of the value function, independent of the chosen value function approximation.

4.1.4.5 Convergence of ADP Algorithms

Let us conclude this introduction to ADP with some brief remarks concerning the convergence of ADP algorithms compared to their counterparts used to solve Markov decision problems. The standard value iteration as well as the standard policy iteration converge to the optimal solution.

The standard value iteration provides an error bound during each iteration. Let ϵ be the maximum difference between the value functions of two successive iterations. Then, it can be shown that the obtained policy differs from the from the value function of the optimal policy by no more than $2 \cdot \epsilon \cdot \beta \cdot (1 - \beta)$ at any state (Kaelbling et al. [Kae+96]).

The standard policy iteration selects during each iteration a better policy. As the selection of the better policy is based on the exact value of the current policy, and because the number of possible policies is restricted, it is guaranteed that the optimal policy is found.

In general, the ADP algorithms do not provide a guarantee that the optimal solution is found. There are some formulations (depending amongst others on the chosen value function representation) that provide a guaranteed convergence, but these formulations require an iteration over the whole state space and are therefore often not very efficient. The algorithm we use to optimize power plant investments does not guarantee that we reach the optimal solution.

4.2 Application of ADP to the Power Generation Expansion Problem

In this section we present how ADP can be used to solve the power generation expansion problem. We start with a description of the general idea, before we discuss several possible state representations. We propose an indexed piecewise-linear value function approximation to determine the value of a state. Thereby, we approximate the value of new units with a piecewise-linear value function. To consider the influence of current fuel and carbon prices as well as the influence of other new units, we define several climate spread and residual demand clusters. The climate spread clusters indicate the cost difference between new CCGT and new coal units, while the residual demand clusters indicate the difference between the electricity demand and the installed capacity excluding the capacity of the considered new units. We use a separate value function approximation for each combination of climate spread and residual demand clusters.

Next, we describe how we determine the marginal value for additional units and how we update the value function approximations. For the update of the value function approximations, we propose two methods, which turn out to be important for our algorithm. The first method can be seen as some sort of hierarchical updating: during the first iterations, we do not only update the value function approximations of the states we visit, but also those of neighboring climate spread and residual demand clusters.

The second method we propose exploits some problem-specific knowledge: if we compare two value function approximations for the same climate spread cluster, but a different residual demand cluster, we know that the value function approximation representing the smaller residual demand must be smaller than or equal to the one representing the greater demand. If an update of a value function approximation violates this relationship, we also update the value function approximations where the violations occur.

Besides these two updating methods, we also propose an problem-adjusted version of the ϵ-greedy exploration. We conclude this section with an overview of the algorithm we use.

4.2.1 General Idea

The SDP formulation of the power generation expansion problem suffers from the curse of dimensionality. The state space gets too large, and we have to use several simplifications. The huge state space leads to two different problems, which might be avoided using ADP: First, due to memory restrictions, we cannot save all states in a look-up table. Value function approximations used in ADP can significantly reduce the memory requirements.

Second, the determination of the contribution margins for all states can be computationally intractable. Applying an ADP algorithm, we do not iterate over the whole state space. Consequently, we do not need to determine the contribution margins for all states.

In this subsection, we motivate our choice of a piecewise-linear concave value function approximation. We start with a simplified deterministic example, before we consider stochastic fuel and carbon prices and the influence of other new units.

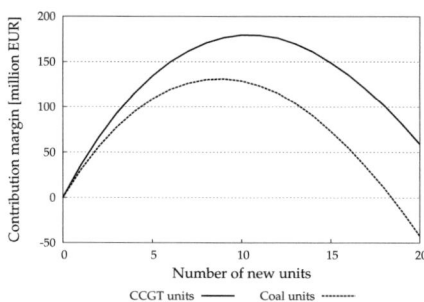

Figure 4.2: Contribution margin depending on the number of new units

4.2.1.1 Deterministic Fuel Prices

Let us start with a simplified example to illustrate how ADP can be used to model the power generation expansion problem. Assume that fuel and carbon prices are known with certainty. Obviously, the profit of the considered operator depends on the number of new units.

Figure 4.2 illustrates this relationship. It shows an example for the contribution margins of new CCGT units and new coal units during one year as a function of the number of new units. If there are only a few new units, the (total) contribution margin increases, if an additional unit is added. However, if there are more than ten CCGT units or nine coal units, an additional unit decreases the contribution margin in this example.

If we add an additional new unit to the existing units, then the contribution margin for this additional new unit is equal to or less than the contribution margin of a new unit without this additional unit. Let us illustrate this effect by comparing two scenarios: in scenario S^n, we have n new units (of the same type, e.g., CCGT units), while in scenario S^{n+1}, we add an additional unit.

The contribution of a new unit is equal to the product of the amount of generated electricity and the difference between the electricity price and the variable costs of the unit. Let us consider how these three factors change when an additional unit is added:

- The variable costs of a unit are independent of the number of new units. Hence, the variable costs do not change.

- The amount of electricity that the additional new unit produces must be equal to or lower than the amount of electricity produced by a new unit in scenario S^n. The reason is that the unit commitment is determined by cost minimization. As the variable costs of the additional new unit are the same as the variable costs of the other new units, the electricity production of the additional new unit cannot be greater than the electricity production of a new unit in scenario S^n if the two scenarios differ only in the number of new units.[3]

- The electricity price is set by the most expensive unit producing electricity. The only difference between scenario S^n and S^{n+1} is the additional new unit. If electricity demand is lower than the cumulated capacity of all units that have lower variable costs than the considered new units, the electricity price remains the same. Otherwise, the new unit replaces the electricity production of a more expensive unit. Hence, the electricity price in scenario S^{n+1} is lower than or equal to the one in scenario S^n.

So far we have only considered the contribution margin of the additional new unit. However, an additional new unit has also an impact on the contribution margins of the existing units, as it might result in a lower electricity production per new unit and a lower electricity price.

Let us consider instead of the contribution margin of the additional new unit the cumulated contribution margin of all new units. If we add an additional new unit, the variable costs remain the same. The cumulated electricity production of all new units either increases or it remains constant. The electricity price will either be the same or decrease.

The opposed influence of a new unit on the cumulated electricity production and the electricity price leads to the curves depicted in Figure 4.2. For the first additional unit, the increase in electricity production outweighs the decrease of the electricity price, and the cumulated contribution margin increases. If more units are added, the effect of the decreasing electricity price prevails. For the example shown in Figure 4.2, the contribution margins are a concave function of the number of new units. Hence, we might use a piecewise-linear concave function to approximate the contribution margins depending on the number of new units.

However, we have no guarantee that the contribution margins are a concave function of the number of new units. A counter-example can be easily constructed, as we see in the following:

---

[3]We assume here that units with the same variable costs (and the same technical restrictions) produce the same amount of electricity during each hour. This might not be the case if a unit with variable costs c is the marginal unit, i.e., the remaining unsatisfied electricity demand is lower than the capacity offered by all units with variable costs c. In such a case, we assume that the remaining unsatisfied demand is evenly distributed amongst all units with variable costs c.

Example. *Let the capacity of a new unit be 1,000 MW and the variable costs 20 €/MWh. Assume that the electricity price is 100 €/MWh if we add one new unit, 75 €/MWh if we add two new units, and 70 €/MWh in case of three new units. Assume further that the new units produce at their maximum capacity in all three cases.*
The resulting contribution margins per hour are 80,000 € in case of one unit, 110,000 € if two units are added, and 150,000 € if three units are added. The resulting function is non-concave.

Despite this counter-example, tests based on our LDC electricity market model using the existing merit–order curve indicate that the contribution margins of additional units—considered over a whole year—can be approximated relatively well with a piecewise-linear concave function. Even if the value function is non-concave, the ADP approach based on piecewise-linear value function approximations might work following Powell [Pow07, p. 353]: *They* [linear value function approximations] *can also work well in settings where the value function increases or decreases monotonically, even if the value function is neither convex nor concave, nor even continuous.*

4.2.1.2 Stochastic Fuel and Carbon Prices

So far we made the simplifying assumption that fuel and carbon prices are known with certainty. Obviously, this is not the case. As fuel and carbon prices have an important influence on the contribution margins, we must consider fuel and carbon price uncertainty. There is a functional relationship between fuel and carbon prices and the contribution margins. However, this relationship is non-linear and difficult to capture by function approximations.

Hence, we do not approximate the influence of stochastic fuel and carbon prices with a function approximation. Instead, we use the indexed piecewise-linear function approximation approach for our problem. Thereby, we define different fuel and carbon price scenarios. For each of these scenarios, we use a piecewise-linear value function approximation to estimate the value of CCGT units and coal units. The question remains how to define these fuel and carbon price scenarios. Similar to the normal SDP approach, there is a tradeoff between accuracy and running time for the number of considered scenarios. The more scenarios we consider, the more value functions must be approximated.

Based on the same reasoning as the one used for the ISDP model, we use the climate spread instead of fuel and carbon prices to represent the fuel and carbon price scenarios. We define several climate spread scenarios, where each of these climate spread scenarios represents an interval of climate spreads. For each of the climate spread scenarios, we use a piecewise-linear value function approximation to estimate the value of new units.

4.2.1.3 Dependence on Other Units

Another implicit assumption we have made so far is that the contribution margins of the different types of new units are independent of each other. This assumption is implicitly made, as the approximated values for the different attributes (in our case the number of new units of the different types) are summed up.

In reality, the contribution margins of CCGT units are not independent of the number of coal units (and vice versa). Additional units of a different technology have a similar impact on the contribution margins as units of the same type: they can change the electricity production of the considered units as well as the electricity price. As a consequence, we cannot neglect this influence.

To handle this problem, we add an extra attribute for each considered type of new unit to our states: the *residual electricity demand*. We define the residual electricity demand as the electricity demand less the available capacity of all units except the capacity of new units of the considered type. For example, the residual electricity demand for new coal units is equal to the electricity demand minus the total available capacity of all units except the capacity of new coal units built by the considered operator. For the determination of the residual electricity demand, we take the maximum annual demand.

As for the climate spread scenarios, we define different intervals for the residual electricity demand scenarios. For each of these scenarios, we use a separate piecewise-linear value function approximation for the number of new units of each type.

4.2.1.4 Overview

Figure 4.3 illustrates the ideas presented so far in this subsection. It shows the contribution margins as a function of the number of new units for different climate spread and residual demand clusters. With our ADP algorithm, we try to approximate each of these contribution margins with a piecewise-linear concave function approximation.

There is one relationship between the different value functions that we are going to exploit during the updates of the value function approximations. Consider the

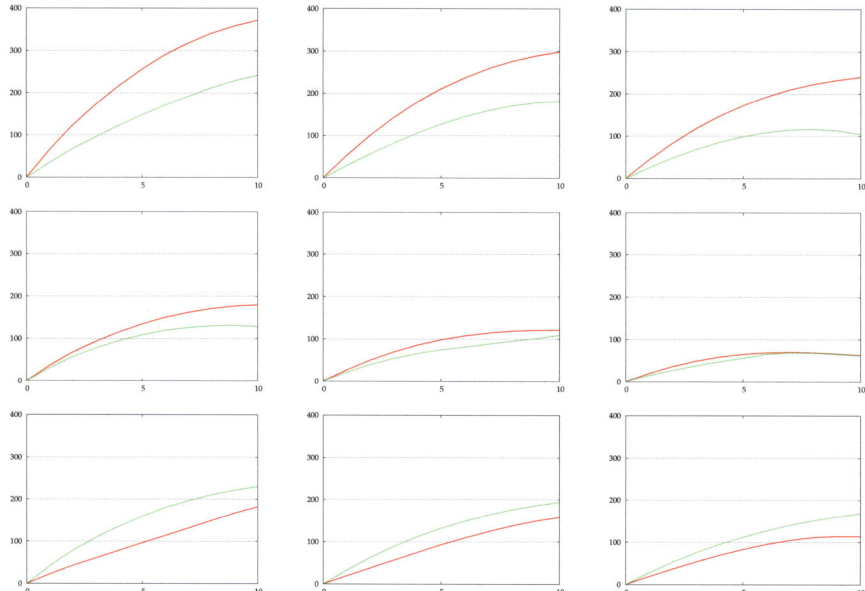

Figure 4.3: Contribution margins as a function of the number of new units for CCGT units (red) and coal units (green). On the x-axis, the number of new units is depicted, on the y-axis the cumulated contribution margin. In each row, a different climate spread is used, each column represents another residual demand scenario.

different contribution margins in the same row. For each row, the climate spread cluster is the same, only the residual demand cluster differs. The residual demand decreases from the left to the right. With a decreasing residual demand, also the contribution margins decrease. As we use the same fuel prices for the figures depicted in the same row, the costs of electricity production of the new units remain constant. A decreasing residual demand results in a constant or decreasing electricity production and a constant or decreasing electricity price. Hence, the contribution margins of the new units are either constant or they decrease. We describe in Subsection 4.2.4.2 how we exploit this relationship.

4.2.2 State Representation

In the previous part of this section, we outlined the general idea how ADP can be applied to our problem. We presented an approach based on indexed, piecewise-linear value function approximations. However, we have been a bit vague about the state representation and how the value of a state is approximated. Before we start describing our state representation, we would like to stress that an adequate modeling of the considered problem is extremely important and—to the author's experience—more difficult than in normal DP.

In the following, we first describe a straightforward state representation based on the ideas presented in the previous part of this section. Unfortunately, this representation suffers from a problem that is not obvious at first sight: it cannot capture the option value of postponing an investment. For that reason, we present an alternative state representation that is able to consider this option value. In the last part of this subsection, we describe how investments of competitors can be considered.

4.2.2.1   Extensive State Representation

A straightforward implementation of the ideas presented before would represent a state by the number of new units of each type during each year and the current fuel prices. Assuming that we model the coal price, the gas price and the carbon price as stochastic processes, a state would have the following representation:

$$x_t = \begin{pmatrix} x_t^{CCGT} \\ \vdots \\ x_T^{CCGT} \\ x_t^{Coal} \\ \vdots \\ x_T^{Coal} \\ c_t^{Coal} \\ c_t^{Gas} \\ c_t^{CO_2} \end{pmatrix},$$

where x_t^{CCGT} denotes the number of CCGT units in year t and x_t^{Coal} the number of coal units in year t.

Based on this state representation, we can use a piecewise-linear value function approximation for each type of unit and each considered year depending on the climate spread and residual demand cluster. Assume that the information about the current climate spread cluster and the residual demand cluster for unit type u in year t' is represented by $\phi^{ut'}(x_t)$, then we use the following value function

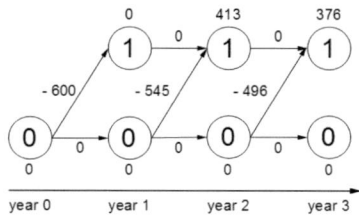

Figure 4.4: Simple example to demonstrate the problem with the extensive state representation.

approximation for state x_t:

$$\tilde{V}_t(x_t) = \sum_{t'=t}^{T} \left(\tilde{V}_{t,\text{CCGT}_{t'}} \left(x_{t'}^{\text{CCGT}} \mid \phi^{\text{CCGT}_{t'}}(x_t) \right) + \tilde{V}_{t,\text{Coal}_{t'}} \left(x_{t'}^{\text{Coal}} \mid \phi^{\text{Coal}_{t'}}(x_t) \right) \right).$$

$$(4.6)$$

Thereby, each $\tilde{V}_{t,u_{t'}}(\cdot)$ on the right hand side of equation (4.6) is a piecewise-linear value function approximation. To determine the value of a state, we simply sum up the value of all attributes representing the number of new units.

However, the presented representation suffers from one substantial drawback: it is not able to capture the option value of delaying an investment. Instead, an investment is carried out, as soon as it gets a positive net present value. Let us visualize this problem with a simple example.

Example. *Assume that we have the possibility to invest in one single unit. The construction time of the unit is one year, and the unit costs 600 million €. To simplify matters, we assume that there is no fuel price uncertainty. The returns of this unit are 0 € in the first year (this could be due to an overcapacity in the market or due to an unfavorable fuel price), 500 million € in the second and third year. We assume a discount rate of 10%. Figure 4.4 visualizes this simple example.*

The optimal solution is obvious. We invest after the first period, this yields a net present value of 244 million €.

Now consider what happens if we apply an ADP algorithm based on the outlined state representation and value function approximation. The marginal values for having an additional unit is 0 €, 413 million € and 376 million € for the years one to three. Starting in the initial state in year 0, we might either invest or not. As the investment costs of 600 million € are lower than the marginal value of a unit in the following years (789 million €), the investment is carried out. This result corresponds to the optimal solution of a traditional net present value calculation applied in year 0. The problem is that the value of being in state 0 in year one is underestimated.

4.2.2.2 Compact State Representation

Let us now consider an alternative state representation, which is similar to the representation chosen for the SDP model. Instead of saving the number of active units for each year of the optimization horizon, we might just save the number of units that are already active plus those that are under construction at the moment:

$$x_t = \begin{pmatrix} x_t^{\text{CCGT}} \\ \vdots \\ x_{t+t^{\text{Constr,CCGT}}}^{\text{CCGT}} \\ x_t^{\text{Coal}} \\ \vdots \\ x_{t+t^{\text{Constr,Coal}}}^{\text{Coal}} \\ c_t^{\text{Coal}} \\ c_t^{\text{Gas}} \\ c_t^{\text{CO}_2} \end{pmatrix},$$

Thereby, $t^{\text{Constr,CCGT}}$ and $t^{\text{Constr,Coal}}$ denote the construction time for new CCGT and new coal units, respectively.

While the state representation looks similar to the representation presented before, the value function approximation works different. At first glance, the value function approximation seems to be almost the same as before:

$$\tilde{V}_t(x_t) = \sum_{t'=t}^{t+t^{\text{Constr,CCGT}}} \tilde{V}_{t,\text{CCGT}_{t'}} \left(x_{t'}^{\text{CCGT}} \mid \phi^{\text{CCGT}_{t'}}(x_t) \right)$$
$$+ \sum_{t'=t}^{t+t^{\text{Constr,Coal}}} \tilde{V}_{t,\text{Coal}_{t'}} \left(x_{t'}^{\text{Coal}} \mid \phi^{\text{Coal}_{t'}}(x_t) \right).$$

As before, the approximations for $\tilde{V}_t(x_{t'})$ with $t' < (t + t^{\text{Constr}})$ are piecewise-linear value functions approximating the contribution margin in the corresponding year depending on the number of new units.

The difference to the value function approximation used before is the approximation used for $\tilde{V}_{t,\text{CCGT}_{t'}} \left(x_{t'}^{\text{CCGT}} \mid \phi^{\text{CCGT}_{t'}}(x_t) \right)$ with $t' = t + t^{\text{Constr,CCGT}}$, and the one used for $\tilde{V}_{t,\text{Coal}_{\hat{t}}} \left(x_{\hat{t}}^{\text{Coal}} \mid \phi^{\text{Coal}_{\hat{t}}}(x_t) \right)$, where $\hat{t} = t + t^{\text{Constr,Coal}}$. As we do not explicitly consider the number of new units until the end of the optimization horizon, these value functions are used to approximate the value of having the corresponding number of units from year $t + t^{\text{Constr}}$ until T. This value does not necessarily correspond to the cumulated contribution margin for the years from $t + t^{\text{Constr}}$ until T. We will demonstrate this in Subsection 4.2.3, where we discuss how the slopes of the value function approximations are calculated. Then, we also show that this compact representation avoids the problem of the non-consideration of the option value to postpone an investment, which we encountered with the extensive representation.

4.2.2.3   Investments of Competitors

If we model investments of competitors, we have to add additional attributes to a state counting the number of new units built by competitors and those under construction. For the compact state representation, we add the attributes $x_t^{\text{CCGT,Ext}}, \ldots, x_{t+t^{\text{Constr,CCGT}}}^{\text{CCGT,Ext}}, x_t^{\text{Coal,Ext}}, \ldots, x_{t+t^{\text{Constr,Coal}}}^{\text{Coal,Ext}}$.

While the representation of the new units built by competitors corresponds to the one of the units of the considered operator, the way they are used to approximate the value of a state is different. For each of the attributes representing the number of units built by the considered operator, we use a piecewise-linear value function to determine the value of these attributes. New units built by competitors decrease the value of the units built by the considered operator as they tend to reduce the electricity production of other units and to decrease electricity prices.

We account for this effect with the residual demand clusters. The construction of new units by competitors reduces the residual demand for the units of the considered operator. Hence, if competitors build new units, we use a value function approximation for the number of own units that is smaller or equal to the one used if no external investments are carried out.

4.2.2.4   Climate Spread and Residual Demand Clusters

During the presentation of the state representations we focused so far on the attributes used for the piecewise-linear value function approximations. We have not specified how we define the climate spread and the residual demand clusters, which we just denoted with $\phi^{u_{t'}}(x_t)$.

Climate Spread Clusters

Let us start with the climate spread clusters. For each year, we define a fixed number of climate spread clusters. Each climate spread cluster covers a predefined interval of climate spreads. These intervals may change from year to year.

There are two different approaches how these intervals can be determined. We might either chose intervals of the same size, or we might select the intervals such that the probability for each interval is equal. Using intervals of equal size ensures that intervals do not get too large (if enough intervals are chosen). Therefore, the approximation quality of the value function approximations tends to be similar for all intervals.

The second approach results in intervals of different sizes. On the one hand, the intervals around the most probable climate spread scenarios are the smallest. Hence, we can expect the best approximation quality for these intervals. On the other hand, intervals representing less probable climate spread scenarios can get relatively large, and may result in a poor approximation quality. As we think that a good approximation of the most probable scenarios is more important, we opt for the latter approach.

Residual Demand Clusters

As for the climate spread clusters, we define for every year of the optimization horizon a fixed number of residual demand clusters. These intervals may also change from year to year.

Though, there is one important difference between the climate spread clusters and the residual demand clusters. When we approximate the value of a state, we use the same climate spread cluster for all value function approximations, but we may use different residual demand clusters.

We use the current climate spread cluster also for future years, because we do not know how climate spreads will evolve. The residual demand clusters in our model are only influenced by the number of new units.[4] Due to the construction time of new units, the residual demand clusters are fixed for all attributes except those representing the number of new units until the end of the optimization horizon. Hence, the residual demand clusters are at least partly fixed. Thereby, the residual demand clusters may be different for each attribute.

Example. *Let us illustrate with a simple example how the climate spread clusters and the residual demand clusters are used. Let the maximum electricity demand be 15,000 MW. Assume that there are coal units with a capacity of 8,000 MW and CCGT units with a capacity of 5,000 MW (and no other units).*

---

[4]If we considered also demand uncertainty, this uncertainty would influence the residual demand cluster. But also in this case, we propose to stick to the approach we present here.

Assume further that we use two residual demand clusters for CCGT units and coal units. The first residual demand cluster for CCGT units ($\text{rdc}_1^{\text{CCGT}}$) covers all residual demand values that are greater than or equal to zero, while the second one ($\text{rdc}_2^{\text{CCGT}}$) covers all residual demand values that are negative. The residual demand clusters for coal units are defined accordingly.

Assume that we use two different climate spread clusters, and that the current climate spread lies in climate spread cluster one (csc_1). Let the capacity of a new unit be 1,000 MW. To simplify matters, we assume that a new unit has a construction time of one year. We use the compact state representation in this example.

We are currently in year t, and we want to estimate the value of the following state characterized by two CCGT units in year t, five in year t+1 and one coal unit in year t and two coal units in year t+1. Then, we use the following value function approximations:[5]

$$
\tilde{V}\begin{pmatrix} 2 \\ 5 \\ 1 \\ 2 \end{pmatrix} = \tilde{V}_{t,\text{CCGT}_t}(2 \mid \text{csc}_{1,t}, \text{rdc}_{1,t}^{\text{CCGT}}) + \tilde{V}_{t,\text{CCGT}_{t+1}}(5 \mid \text{csc}_{1,t}, \text{rdc}_{1,t+1}^{\text{CCGT}})
$$

$$
+ \tilde{V}_{t,\text{Coal}_t}(1 \mid \text{csc}_{1,t}, \text{rdc}_{1,t}^{\text{Coal}}) + \tilde{V}_{t,\text{Coal}_{t+1}}(2 \mid \text{csc}_{1,t}, \text{rdc}_{2,t+1}^{\text{Coal}}).
$$

Note that we use the same climate spread cluster for all four piecewise-linear value function approximations. The residual demand clusters differ. In year t, the residual demand for CCGT units is equal to 1,000 (remember that we subtract the capacity of the new coal units in year t). This corresponds to the first residual demand cluster. In year t+1, the residual demand is 0. Hence, the residual demand cluster remains the same. For coal units, the residual demand cluster changes. In year t, we have a residual demand of 0 corresponding to residual demand cluster 1. In the next year, the residual demand (-3,000) lies in residual demand cluster 2.

4.2.3 Marginal Value Calculation

To update the piecewise-linear value function approximations, we have to determine the marginal values for the different attributes. In this subsection, we demonstrate how we calculate these marginal values for the compact state representation.

---

[5]For the sake of simplicity, we drop the fuel prices in the state representation and depict only the attributes representing the number of new units on the left hand side of the equation.

Thereby, we can distinguish three cases, which we first describe, before we provide two examples to illustrate the marginal value calculation.

- *Attributes approximating the value of electricity production in the current year t:*
  For these attributes, the determination of the marginal value is straightforward. We first use our fundamental electricity market model to determine the contribution margins for the current state. To determine the marginal value of an additional unit, we run the electricity market model again with the same parameters, except for the additional unit of the considered type. The difference between the two results is used to update the slope of the corresponding value function approximation.

  Formally written, we determine the marginal value of attribute $x_t^{u'}$, $\hat{v}_{t,u_t'}$ as

$$
\hat{v}_{t,u_t'} = R_t \begin{pmatrix} x_t^{u'}+1 \\ x_t^u \\ c_t^{\text{Coal}} \\ c_t^{\text{Gas}} \\ c_t^{\text{CO}_2} \end{pmatrix} - R_t \begin{pmatrix} x_t^{u'} \\ x_t^u \\ c_t^{\text{Coal}} \\ c_t^{\text{Gas}} \\ c_t^{\text{CO}_2} \end{pmatrix}.
$$

  Thereby, $R_t(\cdot)$ denotes the total reward for the new units depending on the current year and the current fuel prices. This value is calculated by our fundamental electricity market model.

- *Attributes approximating the value of electricity production in future years, where the number of units cannot be changed any more:* This are the years t' where $t < t' < t + t^{\text{Constr}}$ with $t^{\text{Constr},u}$ denoting the construction time for units of technology u. While the number of units is fixed for these years, the climate spread may change.

  When we are in year t, we do not yet know how fuel prices evolve. To update these marginal values, we wait until we are in year t', and use the discounted marginal values of the corresponding attributes observed in year t':

$$
\hat{v}_{t,u_{t'}'} = \beta^{t'-t} \cdot \hat{v}_{t',u_{t'}'}.
$$

- *Attributes approximating the cumulated value of electricity production for units of technology u for the period from t' until T, with $t' = t + t^{\text{Constr},u}$:* The determination of the marginal value for these attributes differs from the marginal value calculation of other attributes. The reason is the following: If we apply the approach described before, we would implicitly assume that the number of units for the considered period (from t' to T) is fixed. However, this is not the case.

Remember that we are updating the value function approximation of a post-decision state in year t. As we are considering a post-decision state, the decision for stage t has already been made. Consequently, the number of new units is fixed until year $t + t^{Constr,u} = t'$. With the decision in year $t + 1$, we can change the number of units for all years after t'.

The marginal value of an additional unit in year t' can be decomposed into two parts: the observed discounted marginal value in year t' and the difference in the discounted expected value for new units from year $t' + 1$ until T of the succeeding optimal post-decision states in year $t + 1$.

The discounted marginal value of an additional unit in year t' is $\beta^{t'-t} \cdot \hat{v}_{t',u'_{t'}}$. The difference in the discounted expected value for new units from year $t' + 1$ until T of the succeeding optimal post-decision states is a bit more difficult to get. Let x_t^a be the post-decision state for which we want to update the value function approximation, and $\overline{x}_t^a$ be a state equal to x_t^a, except for the attribute $u'_{t'}$, where $\overline{x}_t^a$ has one extra unit.

We now simulate one outcome of the stochastic fuel prices, and determine the resulting pre-decision states in year $t + 1$ for both states. We denote these states with x_{t+1} and $\overline{x}_{t+1}$. Next, we determine the optimal decision for both pre-decision states, i.e., we solve line 5 of Algorithm 5 (page 189).

Let the optimal decisions be a_{t+1}^* and $\overline{a}_{t+1}^*$, and the post-decision states that are reached when we apply these optimal decisions x_{t+1}^a and $\overline{x}_{t+1}^a$, respectively. At first glance, it might be straightforward to to use the difference between the expected values of these two states to update the value function approximation.

However, we use a slightly different approach, for the following reason: x_{t+1}^a and $\overline{x}_{t+1}^a$ differ only in the attributes representing the number of units of the considered type u' in year t' and those representing the number of new units of any type u from $t' + 1$ until T. If we took the difference between the expected values of these two states, we would use the expected value of having an additional unit in year t'. However, we do not want to use the expected value, but the observed value. For that purpose, we use the observation $\hat{v}_{t',u'_{t'}}$.

To consider the difference in the expected value of the units from year $t' + 1$ until T, we just calculate the difference between the expected value of these attributes. The marginal value used to update the value function approximation is

$$\hat{v}_{t,u'_{t'}} = \beta^{t'-t} \cdot \hat{v}_{t',u'_{t'}} +$$
$$\beta \cdot \left(\sum_{u} \left(\tilde{V}_{t+1,u_{t'+1}}(\overline{x}^u_{t'+1}) - \tilde{V}_{t+1,u_{t'+1}}(x^u_{t'+1}) \right) - \left(c(\overline{a}^*_{t+1}) - c(a^*_{t+1}) \right) \right).$$
$$(4.7)$$

Thereby, $c(\overline{a}^*_{t+1})$ and $c(a^*_{t+1})$ denote the costs for the decisions $\overline{a}^*_{t+1}$ and a^*_{t+1}, respectively.

Let us now illustrate the determination of the marginal values based on two examples. The first one illustrates the case of $t' < t + t^{\text{Constr}}$, while the second example focuses on the case $t' = t + t^{\text{Constr}}$.

Example. *An investor may invest in (several) CCGT units. Let the construction time of these units be two years. Assume that there are two different climate spread clusters, which we denote with $\text{csc}_{t,0}$ and $\text{csc}_{t,1}$. For the ease of simplicity, we assume a linear value function approximation. As there is only one type of new units, we do not have to consider different residual demand clusters.*

The contribution margin per unit is equal to 0 if fuel prices lie within $\text{csc}_{t,0}$, and 100 if fuel prices are within $\text{csc}_{t,1}$. Assume that we are in year t and fuel prices lie in climate spread cluster $\text{csc}_{t,1}$. Consequently, the marginal value of an additional unit in year t is equal to 100.

The marginal value of an additional unit in $t + 1$ depends on the fuel price development. To update this value, we simulate the fuel prices for $t + 1$. If these fuel prices are in climate spread cluster $\text{csc}_{t+1,0}$, we use 0 for the update, otherwise we use 100.

Assume that the transition probabilities from $\text{csc}_{t,1}$ to $\text{csc}_{t+1,0}$ and to $\text{csc}_{t+1,1}$ are 0.5 each. Then, the value function approximation of $\tilde{V}_{t,CCGT_{t+1}}(x^{a,CCGT}_{t+1} \mid \text{csc}_{t,1})$ will converge to 50. We drop the update of $\tilde{V}_{t,CCGT_{t+2}}(x^{a,CCGT}_{t+2} \mid \text{csc}_{t,1})$ here, as this update mechanism is illustrated by the next example.

Example. *Let us come back to the example on page 210 used to demonstrate the inability of the extensive state representation to capture the option value of postponing an investment.*

t	0	1	2	3
$\tilde{V}_{t,u_t}(1)$	0	0	413	376
$\tilde{V}_{t,u_{t+1}}(1)$	545	789	376	0

Table 4.1: Value function approximations (after convergence) for having one unit.

Let us consider how the compact state representation can be used in this example. As we assume a construction time of one year, a state consists of two attributes: the number of units active in year t, and those active from year $t+1$ until the end of the optimization horizon. Table 4.1 shows the resulting value function approximations (after convergence of the algorithm) for having one unit. The first row represents the value of having one unit in the current year, the second row shows the cumulated value of having one unit from the next year until the end of the optimization horizon.

The values in the first row year correspond to the contribution margins for the electricity production in the corresponding year. Let us demonstrate based on year 0 how the values of the second row are calculated. We assume that the value function approximations of the succeeding years have already converged to the values shown in Table 4.1.

Let the current state be $x_0^a = (0,0)^T$. To determine the marginal value of an additional unit for the years 1 to 3, we use the state $\overline{x}_0^a = (0,1)^T$.

To determine the succeeding pre-decision states, we would normally simulate the fuel prices for the next year now. But as we disregard fuel price uncertainty in this example, we can skip this step. The succeeding pre-decision states are $x_1 = (0,0)^T$ and $\overline{x}_1 = (1,1)^T$. We now determine the optimal decision for both states. As we allow only one new unit, the only decision in state $\overline{x}_1$ is to build no additional new units. The succeeding post-decision state is $\overline{x}_1^a = (1,1)^T$.

The optimal decision for state x_1 is to invest. The expected value of the post-decision state $x_1^a = (0,1)^T$ minus the costs for investing is 244 million € (789 million € − 545 million €), which is greater than the expected value for not investing (0 €).

Now, we have all values we need to determine the marginal value for an additional unit following equation (4.7). In this example, the observed value for an additional unit in year one, $\hat{v}_{1,u_1'}$ is 0 €. The number of new units in the years two to three is one for both x_1^a and $\overline{x}_1^a$. The expected value for these units is 789 million €. The only difference are the costs of the optimal decisions, which are 0 € for $\overline{a}_1^*$ and 545 million € for a_1^*. Hence, the marginal value used to update the value function approximation in year zero is:

$$\hat{v}_{0,u_1} = 0 + (789 - 789) - (0 - 545) = 545 \, [million \, €].$$

As the values used in the example are already discounted values, we have dropped the discount factor in equation (4.7).

When we decide whether to invest or not in year 0, then the investment is postponed, as the investment costs (600 million €) are higher than the expected value of a unit (545 million €).

4.2.4 Value Function Approximation Updates

The marginal values, which are determined as described in the previous sub-section, are used to update the piecewise-linear value function approximations. Thereby, we use a separate approximation for each attribute representing the number of new units depending on the climate spread cluster and the residual demand cluster. The introduction of different climate spread and residual demand clusters allows for a better approximation of the value function, but it also increases the number of value functions approximations, which must be updated. To improve the speed of convergence, we introduce two methods, which we describe in the following.

The first method is based on the idea that it can be beneficial to update not only the value function approximations of the visited state during the first iterations, but also those with similar climate spread clusters and similar residual demand clusters. The second approach exploits some problem-specific knowledge concerning value function approximations with the same climate spread but different residual demand clusters. If all parameters but the residual demand cluster are the same, then the value function approximation for a greater residual demand is in general greater than or equal to the one with a smaller residual demand cluster.

4.2.4.1 Initial Updates of Neighboring Value Function Approximations

One design choice of an ADP algorithm is the initialization of the value function approximations. It is desirable to choose the initial values close to the real values. However, this is difficult because we have no knowledge about these real values at the beginning of the algorithm. As a consequence, we have to start with more or less arbitrarily chosen initial values, and hope that the approximation converges quickly to the real values.

The problem is that the more value function approximations we have, the longer it takes until these functions converge, as we update only a subset of these functions during each iteration. It may take a while, until each function is at least updated

---

Algorithm 7 Selection of neighboring value function approximations

---

1: **if** $i < i^{\max}$ **then**

2:     $d_i^{\text{rel}} = (1 - i/i^{\max}) \cdot d^{\text{rel},\max}$

3:     $d_i^{\text{csc}} = \text{ceil}\big((n^{\text{csc}} - 1) \cdot d_i^{\text{rel}}\big)$

4:     $\text{csc}_i^{\min} = \max\big(0, csc(i) - d_i^{\text{csc}}\big)$

5:     $\text{csc}_i^{\max} = \min\big(n^{\text{csc}} - 1, csc(i) + d_i^{\text{csc}}\big)$

6:     **for** $j = \text{csc}_i^{\min}$ **to** $\text{csc}_i^{\max}$ **do**

7:        $d_{ij}^{\text{csc,rel}} = |(csc_j - csc(i)|/(n^{\text{csc}} - 1)$

8:        $d_{ij}^{\text{rdc}} = \text{ceil}((n^{\text{rdc}} - 1) \cdot \max(0, d_i^{\text{rel}} - d_{ij}^{\text{csc,rel}}))$

9:        $\text{rdc}_{ij}^{\min} = \max\big(0, rdc(i) - d_{ij}^{\text{rdc}}\big)$

10:       $\text{rdc}_{ij}^{\max} = \min\big(n^{\text{rdc}} - 1, rdc(i) + d_{ij}^{\text{rdc}}\big)$

11:       **for** $k = \text{rdc}_{ij}^{\min}$ **to** $\text{rdc}_{ij}^{\max}$ **do**

12:          $\tilde{V}_{t-1,u_{t'}}^{i}(x_{t-1}^{a,u,i}|csc_j, \text{rdc}_k) \leftarrow U^V(\tilde{V}_{t-1,u_{t'}}^{i-1}, x_{t-1}^{u,i}, \hat{v}_{t,u_t'}^{i})$

13:       **end for**

14:    **end for**

15: **else**

16:    $\tilde{V}_{t-1,u_{t'}}^{i}(x_{t-1}^{a,u,i}|csc(i), rdc(i)) \leftarrow U^V(\tilde{V}_{t-1,u_{t'}}^{i-1}, x_{t-1}^{u,i}, \hat{v}_{t,u_{t'}}^{i})$

17: **end if**

---

once. Hence, the algorithm works for several iterations with value function approximations that are still equal to the initial approximations.

While we have no knowledge about the real value function when we start the algorithm, we iteratively gather new information during each iteration. When we get some information about a value function approximation $\tilde{V}_{t,u_{t'}}\left(x_{t'}^{u} \mid \text{csc}_a, \text{rdc}_a\right)$ of climate spread cluster csc_a and residual demand cluster rdc_a, we can assume that value function approximations of the neighboring climate spread cluster csc_b or the neighboring residual demand cluster rdc_b will be similar. At least, we can assume that $\tilde{V}_{t,u_{t'}}\left(x_{t'}^{u} \mid \text{csc}_a, \text{rdc}_a\right)$ is a better approximation of $\tilde{V}_{t,u_{t'}}\left(x_{t'}^{u} \mid \text{csc}_b, \text{rdc}_a\right)$ than the initial values.

Based on this idea, we update also neighboring value function approximations of the current climate spread and residual demand cluster during the first iterations. To select the neighboring value function approximations, we use Algorithm 7.

For this algorithm, we assume that the climate spread clusters as well as the residual demand clusters are sorted and numbered from 0 to $n^{\text{csc}} - 1$ and $n^{\text{rdc}} - 1$, respectively. Thereby, n^{csc} and n^{rdc} denote the number of climate spread clusters and the number of residual demand clusters, respectively. We denote the current iteration of the Monte Carlo value policy iteration with i, and the climate spread

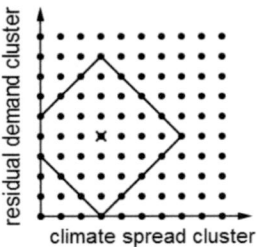

Figure 4.5: Update of neighboring value function approximations.

cluster and the residual demand cluster of the considered value function approximation with $csc(i)$ and $rdc(i)$, respectively. We assume that we update the value function approximation for attribute $u_{t'}$ (e.g., if we use attribute $coal_1$, we update the value function approximation for new coal units in the next year).

Before we can apply the algorithm, we need to specify two values: the number of iterations i^{max} during which the neighboring value function approximations are updated, and the maximum relative distance $d^{rel,max}$ to the current climate spread and residual demand cluster. For example, if we use $d^{rel,max} = 0.5$, and the current climate spread cluster is the one in the middle, we update the value function approximations of all other climate spread clusters (but the same residual demand clusters) during the first iterations.

The algorithm updates neighboring value function approximations as depicted in Figure 4.5. All value function approximations inside (or at the boundary) of the cropped square around the current value function approximation (marked with an x) are updated. With an increasing number of iterations, the square gets smaller, until $i = i^{max}$. Then, we do not update neighboring value function approximations any longer.

4.2.4.2 Update of Other Residual Demand Clusters

The second method we use to speed up convergence is based on the relationship of different value function approximations. Let us consider two scenarios that differ only in the residual demand. Then, we can assume that the contribution margin of new units is at least the same for the scenario with the higher residual demand compared to the other scenario.[6] In the scenario with the higher residual demand, the electricity production of the considered units is at least as high as in

---

[6]This holds as long as the only variable influencing the residual demand is the number of new units of the competing technology. If the demand is also uncertain, or if investments by

the other scenario. The same holds for electricity prices. This knowledge can be exploited when we update the value function approximations.

Let us assume that the residual demand clusters are sorted and rdc_0 is the residual demand cluster with the smallest residual demand. Then, the following relation must hold:

$$\tilde{V}_{t,u_{t'}}(x^u_{t'} \mid csc_{i,t}, rdc^u_{0,t'}) \leq \tilde{V}_{t,u_{t'}}(x^u_{t'} \mid csc_{i,t}, rdc^u_{1,t'}) \leq \ldots$$
$$\leq \tilde{V}_{t,u_{t'}}(x^u_{t'} \mid csc_{i,t}, rdc^u_{n^{rdc}-1,t'}). \quad (4.8)$$

Every time we update a value function approximation, we check whether relation (4.8) still holds. If inequalities (4.8) are violated, we keep the value function approximation of the residual demand cluster that is updated at the moment, and adjust the other value function approximations that violate inequalities (4.8). Assume that an update of the value function approximation belonging to the residual demand cluster 0 results in values that are greater than those of the corresponding value function approximation belonging to residual demand cluster 1. Then, we update the value function approximation belonging to residual demand cluster 1 as follows:

$$\tilde{V}_{t,u_{t'}}(x^u_{t'} \mid csc_{i,t}, rdc^u_{1,t'}) = \max \left(\tilde{V}_{t,u_{t'}}(x^u_{t'} \mid csc_{i,t}, rdc^u_{1,t'}), \tilde{V}_{t,u_{t'}}(x^u_{t'} \mid csc_{i,t}, rdc^u_{0,t'}) \right),$$
$$\forall x^u_{t'} = 0, \ldots, n^u. \quad (4.9)$$

Thereby, n^u denotes the maximum number of new units of technology u. Note that such an update may violate the concavity of $\tilde{V}_{t,u_{t'}}(x^u_{t'} \mid csc_{i,t}, rdc^u_{1,t'})$. If that is the case, we use the convex hull of the function.

Example. *Assume we have* $\tilde{V}_{t\,u_{t'}}(1 \mid csc_{i,t}, rdc^u_{0,t'}) = 100$, $\tilde{V}_{t,u^u_{t'}}(1 \mid csc_{i,t}, rdc^u_{1,t'}) = 110$. *After an update,* $\tilde{V}_{t,u_{t'}}(1 \mid csc_{i,t}, rdc^u_{0,t'})$ *increases to* 115. *Then, we set* $\tilde{V}_{t,u_{t'}}(1 \mid csc_{i,t}, rdc^u_{1,t'}) = 115$.

4.2.5   Modified ϵ-Greedy Exploration

Another problem specific adaption we apply is related to the form of exploration. The normal ϵ-greedy exploration uses the same probability for each decision if an exploration step is taken. For our problem, this is suboptimal. The reason is that

---

competitors are considered, this cannot be guaranteed any longer. Nevertheless, the contribution margin still tends to be higher for the case with a higher residual demand.

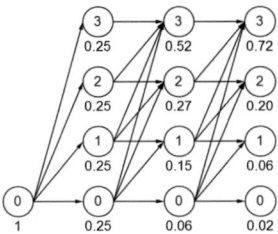

Figure 4.6: Probabilities for different states if ϵ-greedy exploration with $\epsilon = 1$ is applied.

we allow only the construction of new units, but not the shut-down of new units. If we use the same probabilities for each decision, states with many new units are more often visited than states with few new units.

Consider the example depicted in Figure 4.6. Each circle represents a state. The numbers in the circles indicate the number of new units, while the numbers below the states correspond to the occurrence probability of that state, if each decision of a state has the same probability. It can be seen that such an approach favors the exploration of states with many units.

As the objective of an exploration method is to explore different parts of the state space, we propose the following alternative, to which we refer as modified ϵ-greedy exploration. The difference between the modified ϵ-greedy exploration and normal ϵ-greedy exploration is the choice of the probability of each decision if an exploration step is done. Instead of uniform probabilities, we calculate a weight for every decision. This weight is based on the succeeding state that is reached when the decision is applied.

Let $N^n(x_{t,i})$ be the number of time the i-th state at stage t has been visited in iteration n. Then, we iteratively define the weights $W^n(x_{t,i})$ starting at the last stage as

$$W^n(x_{t,i}) = N^n(x_{t,i}) + \frac{\sum\limits_{x_{t+1,i'} \in \mathcal{S}(x_{t,i})} W^n(x_{t+1,i'})}{|\mathcal{S}(x_{t,i})|},$$

where $\mathcal{S}(x_{t,i})$ denotes the set of successors that can be reached from state $x_{t,i}$. To simplify matters, we do not consider the climate spread cluster, i.e., we assume that we stay in the same climate spread cluster. These weights are an indicator how often the corresponding state and its possible successors have already been visited. The higher the weight is, the lower should be the probability to select a decision that leads to this state. Based on these weights, we define the probability to select decision $a_{t,i}$ leading to state $x_{t+1,i}$ as:

$$P(a_{t,i}|x_{t,i}) = \frac{W^{n,max}(x_{t+1} \in \mathcal{S}(x_{t,i})) - W^n(x_{t+1,i})}{\sum\limits_{x_{t+1,i'} \in \mathcal{S}(x_{t,i})} (W^{n,max}(x_{t+1} \in \mathcal{S}(x_{t,i})) - W^n(x_{t+1,i'}))}.$$

Thereby, $W^{n,max}(x_{t+1} \in \mathcal{S}(x_{t,i}))$ denotes the maximum value of W^n of one of the successors of state $x_{t,i}$.

4.2.6   Algorithm

In Subsection 4.1.4, we have briefly presented different algorithms to solve ADP problems. We use the Monte Carlo value policy iteration described on page 201 to solve our problem. In this subsection, we are going to discuss the different parts of the algorithm.

4.2.6.1   Determination of Climate Spread and Residual Demand Cluster

Before we start with the Monte Carlo Value Value Policy Iteration, we have to define the climate spread clusters and the residual demand clusters.

The approach we use to determine the climate spread clusters is similar to the one used to determine the initial climate spread scenarios for our ISDP model. We define the climate spread clusters such that they have (approximately) equal probabilities.

For this purpose, we simulate a fixed number of fuel price paths, e.g., 50,000. For each of the simulated fuel prices, we determine the corresponding climate spread. Then, we choose the bounds of the climate spread clusters for every year such that each cluster comprises the same amount of climate spread scenarios. The number of climate spread clusters may increase with time to account for increasing fuel price uncertainty. Similarly, we may use a different number of climate spread clusters depending on the stochastic process chosen to simulate fuel prices.

The residual electricity demand for new coal units and new CCGT units depends on two factors: electricity demand and electricity supply. In our model we assume that electricity demand is known with certainty. Electricity supply can be divided into a fixed and a variable part. The fixed part are the existing units, the variable part are the new units that may be build. Hence, it is sufficient to concentrate on the new units for the determination of the residual demand clusters.

We choose the residual demand clusters the following way: Let n^{rdc} be the number of residual demand clusters, and $p_{u\text{comp}}^{\max}$ the maximum capacity of new units of the competing technology (e.g., if we determine the residual demand clusters for CCGT units, $p_{u\text{comp}}^{\max}$ is the maximum cumulated capacity of new coal units). Then, we select the bounds of the residual demand clusters such that the i-th cluster covers the interval (of new capacity of the competing technology) of $[(i-1) \cdot p_{u\text{comp}}^{\max} / n^{\text{rdc}}, i \cdot p_{u\text{comp}}^{\max} / n^{\text{rdc}})$.[7]

Note that such a division of the residual demand cluster is only reasonable, if the maximum number of allowed new units is chosen such that this maximum number of new units might (at least almost) be build under some circumstances. For example, if we allow up to 20 new coal units, but due to excess capacities in the market, we never build more than 5 units, then it would be better to use non-equidistant intervals. The problem is that we do not know a priori how many units are built by the optimal solution.

It may be more straightforward to refer to the residual demand clusters as competing new capacity clusters given the way we determine these clusters. However, a possible extension to our model is the introduction of electricity demand uncertainty. In that case, one has to consider the demand uncertainty in the definition of the residual demand clusters. Hence, we use the name residual demand cluster, as it incorporates this idea.

4.2.6.2 Initialization

The influence of the initial values for the value function approximation depends on the degree of exploration used when a decision is chosen. The initialization of the value function is crucial, if an algorithm purely based on exploitation is used, i.e., the algorithm always selects the decision that is optimal based on the current value function approximations. For algorithms with a relative high degree of exploration during the first iterations, the initial values are less important.

In general, it is difficult to guess the right values for the value function approximations. Hence, we might chose the initial values such that they either underestimate or overestimate the real value. We tested both approaches, but the performance of the algorithm is similar in both cases. As the simplest method is to start with value function approximations of 0, we use this approach.

---

[7]To be exact, the given intervals are correct for intervals 1 to $n^{\text{rdc}} - 1$. The n^{rdc}-th interval is $[(n^{\text{rdc}} - 1) \cdot p_{u\text{comp}}^{\max} / n^{\text{rdc}}, p_{u\text{comp}}^{\max}]$.

4.2.6.3 Policy Iteration

The outer loop of the Monte Carlo value policy iteration consists of the policy iteration. During each policy iteration, we keep the value function approximations used to select the decisions within the value iterations constant.

Instead of using a fixed number of policy iterations as written in Algorithm 6, we might also use another stopping criterion. For example, we might run the algorithm until the expected value of the optimal policy does only change marginally during several iterations.

For our tests, we use a fixed number of iterations. Based on numerical test, we choose the number of iterations such that the algorithm converges. The number of iterations required for this convergence depends on the chosen stepsize rule and the degree of exploration.

4.2.6.4 Value Iteration

The main part of the algorithm is the value iteration used to evaluate the value of the fixed policy. Therefore, we discuss the value iteration in detail in the following. We start with the initial (pre-decision) state x_0 and select a decision a_0. While the approach described in Algorithm 6 (line 7) is purely based on exploitation (also referred to as *greedy approach*), we use an approach combining exploitation and exploration like the (modified) ϵ-greedy exploration or the Boltzmann exploration.

Let us briefly demonstrate the selection of the decision based on the greedy approach, which choses the decision that maximizes the expected profits. These profits are composed of two parts: the immediate reward for being in the according pre-decision state less the costs of the decision, and the expected value for the resulting post-decision state.

The immediate reward for being in state x_t is determined by our LDC electricity market model with scarcity mark-ups. Note that we can skip this step for the initial states, where no new units are operable (due to the construction time).

The expected reward for being in the according post-decision state is calculated based on the fixed piecewise-linear value function approximations of the current policy. Before we can calculate the expected value of the post-decision state, we have to determine the current climate spread cluster and—for each attribute representing the number of new units—the residual demand cluster.

If we consider investments of competitors (we can use the approach described in Section 3.3 to determine these external investments), we model these investments directly after the decision of the considered operator, before the next post-decision state is reached. The reason for this choice is that we assume that external investments depend on the decisions of the considered operator.

The selected decision is then used to determine the succeeding post-decision state x_0^a. Next, we sample the coal price, the gas price and the carbon price for year one. These prices are required for the selection of the next pre-decision state x_1.

As we use a double-pass procedure to update the value function, we continue stepping forward in time before making the value function updates. Once we reach the final stage, we step backwards and make the value function updates. Thereby we proceed as described in Subsection 4.2.3.

After the updates are performed, we go back to the initial state and start the next value iteration. The difference between this iteration and the previous one are only the sampled fuel prices. The value function approximation used to determine the optimal decision is still the same, as we use the (fixed) value function approximations of the current policy iteration. After we have done a predefined number of value iterations, we update the fixed value function approximations of the policy iteration, increase the policy iteration counter and start the next value iteration with the updated value function approximations.

4.3 Evaluation

Concerning the evaluation of the results of the ADP algorithm, we face the same problems that we have encountered during the evaluation of the solution of our ISDP approach: we do not know the optimal solution. Hence, it is difficult to evaluate the quality of a solution. Consequently, we use the same approach we have used for the evaluation of the solution quality of the ISDP model: we compare the solution of the ADP model with the solutions of other approaches.

In the following, we first describe the data and the parameters we use for the tests. Next, we evaluate the influence of different design parameters. Thereby, we focus on the problem specific approaches we have introduced: the updates of value function approximations of neighboring residual demand and climate spread clusters, and the exploration method. Afterwards, we compare the solution of the ADP model with those of a deterministic model and those of the ISDP model.

4.3.1   Test Data

The following comparisons are based on the Monte Carlo value policy iteration. We use 2,000 policy iterations with 50 value iterations each. To update the value function approximations, we use a double-pass procedure. We apply the search-then-converge stepsize rule with the parameters $\alpha_0 = 1$, $a = 10$, $b = 1,000$ and $\beta = 0.8$. We start with initial value function approximations of zero. We use the modified ϵ-greedy exploration with $c = 0.8$ to balance the degree of exploration and exploitation. The chosen values have been obtained by numerical tests.

Unless otherwise stated, we consider an optimization horizon of 20 years and allow investments in up to eight CCGT units and eight coal units. The contribution margins for new units are determined by the LDC electricity market model with scarcity mark-ups. We perform tests with fuel prices simulated as correlated GBMs, and others with the logs of fuel prices modeled as correlated trended mean-reversion processes. We consider ten climate spread clusters and nine residual demand clusters with a separate value function approximation for each of these clusters. This choice leads to 90 separate value function approximations per year for each attribute representing the number of new units.

The quality of the final solution is evaluated based on 10,000 fuel price simulations. For these 10,000 fuel price simulations, we use the same stochastic process as the one used for the optimization. However, we initialize the random number generator with a different random seed.

To get an idea how the solution quality evolves during the runtime of the algorithm, we evaluate the solution of every policy iteration based on 1,000 fuel price simulations. While this number of simulations may be too small for an exact evaluation of the solution, tests have shown that 1,000 simulation runs are a good tradeoff between accuracy and running time.

4.3.2   Influence of Parameter Choices

The performance of an ADP algorithm depends on many different design decisions. Important design decisions are the choice of the state representation, the value function approximation, the stepsizes used for the updates of the value function approximations, and the ratio between exploration and exploitation. In addition, the exploitation of problem specific knowledge can significantly speed up an algorithm.

Often, the choice of one of these design parameters has an impact on the influence of another parameter. For example, the degree of exploration influences how quickly stepsizes may decrease. The evaluation of each of these parameters depending on the choice of the other parameters would be beyond the scope of this thesis.

In this subsection, we focus on the influence of the exploitation of problem specific knowledge. Thereby, we evaluate the influence of the problem specific methods we have introduced: the initial updates of neighboring value function approximations, the updates of the value function approximations with the same climate spread cluster but a different residual demand cluster, and the modified ϵ-greedy exploration.

4.3.2.1 Updates of Neighboring Value Function Approximations

Let us start with the evaluation of the two problem specific update mechanisms. Figure 4.7 shows the solution quality depending on the (policy) iteration for an ADP problem using three different update mechanisms. The first approach, denoted with *No extra updates*, updates only the value function approximations of the visited states. The second approach, denoted with *RDC updates*, updates additionally value function approximations with the same climate spread cluster but a different residual demand cluster as described in Section 4.2.4.2. The third approach (*RDC + neighboring upd.*) updates during the first iterations also value function approximations of neighboring climate spread and residual demand clusters (see Section 4.2.4.1). For this purpose, we use the values of $i^{max} = 2,000$ and $d^{rel,max} = 0.5$. With these values, neighboring value function approximations are updated during the first 40 policy iterations. For comparison purposes, we also perform a test with an algorithm consisting of only residual demand cluster. This approach is denoted with *1 RDC* and does only update the value function approximations of the visited states.

It is striking that the approach updating only the visited states does not work well. During the first 1,000 iterations, the evaluated value of the solution is even negative. During the first iterations, the behavior of this algorithm is similar to a random search. The reason is the following: As we update only the value function approximations of the residual demand clusters and climate spread clusters of the states we visit, many value function approximations remain at their initial values during the first iterations. If the solution we choose during an iteration results in losses, the value function approximations are updated with negative values.

Hence, during the next iteration, the algorithm tends to select another solution with states belonging to residual demand clusters that have not yet been updated. With an increasing number of iterations, the algorithm updates more and more value function approximations and is able to make better decisions.

The problem is that the algorithm makes suboptimal decisions during many iterations, and uses the resulting suboptimal solutions to update the value function approximations. There are two issues with this: First, we may get stuck in a suboptimal solution as we use wrong value function approximations to make our decisions. Second, even if this is not the case, the stepsize used to update the value function approximation may have decreased too fast to correct wrong value function approximations.

There are different possibilities to avoid this problem. The simplest one is to use only one residual demand cluster. The alternative is to stick with the nine residual demand clusters, and to update the value function approximations of the neighboring residual demand clusters and climate spread clusters. As it can be seen in Figure 4.7, each of these three approaches works significantly better. All in all, the *RDC + neighboring upd.* approach works best. Compared to the other approaches, it finds a better solution, and it needs less time to find a good solution. It is a bit surprising to see that the approach with one residual demand cluster performs better than the *RDC updates* approach with nine residual demand clusters.

4.3.2.2   Exploration Method

Another important design decision is the choice of the exploration method. Figure 4.8 shows the results for three different methods: ϵ-greedy exploration, modified ϵ-greedy exploration and Boltzmann exploration. For both ϵ-greedy methods, we calculate the probability for an exploration step as $0.8/\sqrt{N^n(x_t)}$ with $N^n(x_t)$ denoting the number of times state x_t has been visited up to iteration n. The temperature used for the Boltzmann exploration is determined following equation (4.4). Only the first 100 iterations are depicted, as the solution quality changes only slightly afterwards.

It can be seen that Boltzmann exploration does not work well in our problem. This is a general observation we have made considering different test instances. The problem might be due to a bad combination of the exploration method and the stepsizes. As we have tested different stepsizes, the problem seems to be related to the form of exploration.

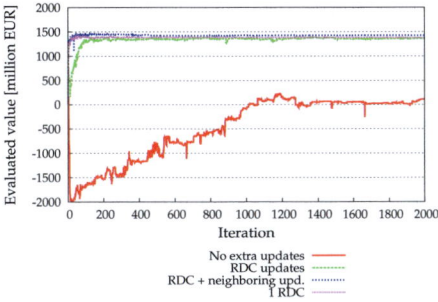
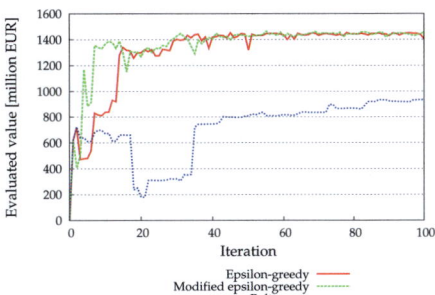

Figure 4.7: Solution quality for different update mechanisms.

Figure 4.8: Solution quality for different exploration methods.

During the first iterations, the modified ϵ-greedy exploration performs better than
the normal ϵ-greedy exploration. This is due to our problem structure. As we
allow only the construction of new units, but not the shut-down, the normal ϵ-
greedy exploration results in higher probabilities for states with many new units
compared to states with few new units during the last years of the optimization
period. The modified ϵ-greedy exploration accounts for this fact and allows a
more uniform exploration of the state space.

For the depicted problem, the advantage of this approach lies in a faster convergence. After approximately 20 iterations, both methods perform similarly.
However, we have also seen test instances where the final solution of the normal ϵ-greedy exploration is slightly worse than the one found with the modified
ϵ-greedy exploration. A possible explanation for such a behavior is a too fast decrease of the stepsizes used to update the value function approximations. Note
that a longer optimization horizon tends to aggravate the problems of normal ϵ-
greedy exploration. This is because the longer the optimization horizon, the more
biased is the probability to explore the states at later stages.

4.3.3   Comparison with Alternative Approaches

In this subsection we compare the solutions of the ADP model with those of a deterministic approach based on expected fuel prices and the solutions of our ISDP
model. For this purpose, we perform tests with different stochastic processes. We
first show the results of these tests, before we discuss the differences between the
ISDP approach and the ADP approach in general.

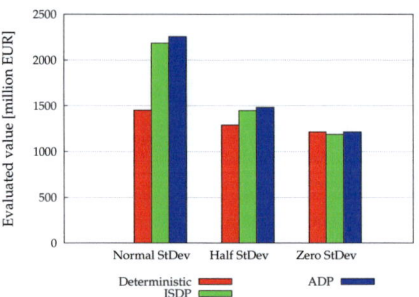

Figure 4.9: Solution quality for different degrees of uncertainty.

4.3.3.1 Results

Figure 4.9 shows the solution quality for one of these test instances with fuel prices simulated as correlated GBMs. We use different standard deviations to evaluate the influence of uncertainty.[8] The optimization horizon for this test is 24 years.

The higher the degree of uncertainty, the greater the advantage of a stochastic approach compared to a deterministic one. The solution of the ADP model performs slightly better than the one of the ISDP model for all three standard deviations shown in Figure 4.9. We have observed similar results for other tests performed with logs of fuel prices modeled as correlated trended mean-reverting processes. We depict the results of the test with fuel prices modeled as correlated GBMs, as the differences in the results are more noticeable as for the tests where the correlated mean-reversion process has been used.

The reason for the better performance of the ADP model is its ability to consider annual decisions. Hence, the considered solution space is greater, and—as the solution space of the ISDP model is a proper subset of the solution space of the ADP model—the optimal solution of the ADP model must be at least as good as the optimal solution of the ISDP model.[9]

In case of deterministic fuel prices, the ISDP model cannot reach the optimal solution due to its inability to consider annual investments. In the tests we have performed, the ADP model finds the optimal solution. However, it must be stressed

---

[8]The standard deviations are the same as the one used for the comparison of the ISDP and the deterministic model on page 173. As we have changed some other parameters (e.g., the optimization horizon), the results are not the same.

[9]We have no guarantee to find the optimal solution. If the quality of the solution of the ADP model is worse than the one of the ISDP model, then we know that the solution of the ADP algorithm is suboptimal. However, if it is the other way around like the results shown in Figure 4.9, we cannot deduce that the solution of the ISDP model is suboptimal within the model.

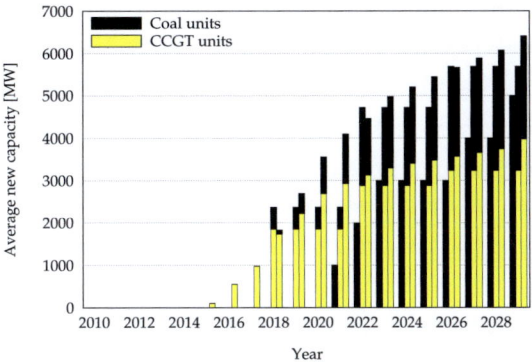

Figure 4.10: Average amount of new capacity built by the solutions deterministic approach (first column), the ISDP model (second column) and the ADP model (third column)

that even in the deterministic case, we have no guarantee that the solution of the ADP model is optimal. For example, if we use the ADP algorithm without updating the value function approximations of other residual demand clusters in the deterministic case, the best solution the algorithm finds results in profits of 1.022 billion €. This is over 15% below the value of the optimal solution, which is 1.214 billion €.

Figure 4.10 shows the average amount of newly installed capacity of CCGT units and coal units for the three different approaches. While the optimal deterministic solution builds only some coal units, the ADP model and the ISDP model start earlier with the construction of new units and prefer CCGT units. The solutions of the ADP and the ISDP model are similar. The possibility of annual investments results in a more uniformly distributed construction of new units of the ADP model. Depending on the realized fuel prices, the construction of new units is already started at the beginning of the second year (2011), while the ISDP model starts with the construction of new units in 2014. On average, the total amount of newly constructed capacity is slightly higher with the ADP model. The difference is mainly due to a higher number of CCGT units.

4.3.3.2 General Comparison with ISDP

The results presented in the first part of this subsection show that the ADP approach delivers better solutions for the considered problems as the ISDP approach.

One may conclude that the ADP method is superior to the ISDP method. However, we do not think that the results allow such a conclusion. Both approaches have their own strengths and weaknesses, which are discussed in the following.

The main advantage of the ADP model is its efficiency. The compact representation of the value function approximation allows us to consider annual investments. Furthermore, we do not need to save a transition matrix for the stochastic fuel prices. For these reasons, the ADP model scales much better than the ISDP model.

As an update of the value function approximation of a visited state does not only apply to this state, but also to other states, the ADP approach finds good solutions relatively fast.

However, there are many drawbacks of the ADP approach. The performance of the ADP model depends strongly on the appropriate choice of the design parameters. Unfortunately, this choice is very problem specific. The consequences of a bad choice can be seen in Figure 4.7 and Figure 4.8: the algorithm delivers very bad results. Note that the possible wrong choices of design parameters are not restricted to these two examples. Also the choice of a inappropriate stepsize rule, or just the wrong choice of a parameter of the selected stepsize rule, can lead to similar problems.

A further difficulty is to detect such problems. As we do not know the optimal solution, we need additional results to which we can compare the quality of our solution. It is obvious that the algorithm denoted with *No extra updates* in Figure 4.7 does not work as its solutions have a negative value. However, if we consider only the Boltzmann exploration in Figure 4.8, then we do not know that the optimal solution is at least 50% better than the one found by the algorithm.

Another drawback of the ADP approach is the assumption of independence of the different attributes for which a function approximation is used. As we have discussed, the profit of new CCGT units is not independent of the number of new coal units. For this reason, we have introduced different residual demand clusters with separate value function approximations for each cluster. The introduction of different residual demand clusters has helped us to improve the solution quality of the algorithm. In the meantime, the different residual demand clusters make the algorithm more complicated. If the normal updating mechanism is not adapted, the algorithm does not work as it could be seen in Figure 4.7.

Our tests have shown that we can use an ADP model to solve the power generation expansion problem with two investment alternatives (CCGT units and coal units). However, it is difficult to integrate either new units of another technology or units of the same technology with different technological constraints (e.g., a different efficiency) into the ADP model.

Assume that we want to consider investments in two different types of coal units and in new CCGT units. Let coal units of type A be more efficient than coal units of type B, and let both of them have lower variable costs (for the current fuel prices) than CCGT units. Consider the value function approximations for coal units B. For the same residual demand cluster, we may get completely different value function approximations. The problem is that we can be in the same residual demand cluster for coal units of type B, when either x of the cheaper coal units of type A or x of the more expensive CCGT units have been build. In the first case, electricity production of units of type B as well as electricity prices tend to be lower than in the second case. The reason is that the position in the merit-order curve of units of type B is different in both cases (as well as the merit-order curve itself).[10]

Compared to the ADP approach, the ISDP approach is less sensitive to suboptimal parameter choices. While there are also some design decisions influencing the performance of the ISDP model (e.g., the splitting criterion), a suboptimal decision has in general less impact than similar choices of the ADP algorithm. For example, if we choose a too low number of splits per iteration in our ISDP model, it takes longer until the algorithm converges. If we choose a stepsize that decreases too fast, this may result in a significantly worse final solution of the ADP model.

Another advantage of the ISDP model is that it is not based on assumptions like the independence of the profits of different types of new units. As a consequence, the ISDP model can consider investments in similar units.

The main disadvantage of the ISDP model is that it does not scale well. To be able to solve the model, we need to introduce several simplifications. One of these simplifications is the temporal aggregation. As a consequence, new units can only be build every four years. Despite the application of temporal aggregation and the introduction of the adaptive grid method, we are restricted to low dimensional problems.

---

[10]Note that it is not sufficient to define the residual demand cluster such that we subtract only the capacity that can produce cheaper than the considered unit from electricity demand. The reason is that the new units of the more expensive technology—in the given example the CCGT units—also influence the electricity price.

4.4   Summary

The main disadvantage of the ISDP model proposed in the previous chapter is its
inability to handle annual decisions. In this chapter, we have presented an alternative method based on ADP to solve the power generation expansion problem.
This problem formulation allows annual investment decisions. To the best of the
author's knowledge, this is the first time that ADP is used to solve the power
generation expansion problem.

In the first part of this chapter, we have given an introduction to ADP. We pointed
out to the differences and similarities between ADP and SDP. As in SDP, there is
a tradeoff in ADP between the immediate reward of a decision and the expected
future reward for the succeeding state based on the chosen decision. While SDP
solves this problem based on the exact expected value of the succeeding state,
ADP uses an approximation of the expected value.

The advantage of using an approximated value is that it does not require a complete enumeration of the state space. The introduction of the post-decision variable, which splits the states into pre-decision states and post-decision states, allows us to replace the standard Bellman equation (4.2), which contains a stochastic
part (the succeeding states depending on the outcome of the stochastic parameter),
with the deterministic optimization problem (4.3). The expectation of the value of
the succeeding states of the original Bellman equation is thereby captured by the
value function approximation of the post-decision state.

Another important difference between SDP and ADP is the way value functions
are saved. In general, SDP saves the value of each state in a lookup-table. For
higher dimensional problems, this becomes computational intractable. While ADP
may also use a lookup-table approach, one of its main advantages is that it allows
more compact forms of value function approximations. Often, the value of a state
can be described as a function depending on some of the attributes of a state. We
have presented two commonly used approaches: linear value function approximations and piecewise-linear value function approximations. In addition, we
presented the indexed (piecewise) linear approximation approach, a combination
of the lookup-table and a function approximation approach. This approach can be
used to approximate the value of states where the value of some attributes can be
approximated by a function, but the value of other important attributes cannot be
captured by a function approximation.

The simplifications introduced by ADP allow it to tackle high-dimensional problems that cannot be solved by SDP. However, these simplifications also have their drawbacks. In general, there is no guarantee that the optimal solution is found. Furthermore, an appropriate choice for the parameters of the algorithm, e.g., the stepsizes used to update the value function approximation or the exploration strategy, is very problem specific. A bad choice of one of these parameters can result in an algorithm that does not work at all.

In the second section of this chapter, we have shown how ADP can be used to model the power generation expansion problem. Based on the observation that the marginal value of an additional power plant is in general diminishing, we use piecewise-linear concave value function approximations for each type of new units. However, the value of a new unit depends also on other factors. These factors are especially the fuel and carbon prices, electricity demand and the power generation fleet in the market. To consider these factors, we use the indexed piecewise-linear value function approximation approach. We define different climate spread and residual demand clusters. The climate spread clusters are used to capture the influence of different fuel and carbon prices, while the residual demand clusters indicate by how much electricity demand exceeds electricity supply. For each combination of climate spread and residual demand cluster, we use a separate piecewise-linear value function approximation to estimate the value of new units.

The introduction of these climate spread and residual demand clusters allows us to consider additional information, but it also complicates the update process of the value function approximations. To speed up the convergence of the algorithm, we have introduced two problem-specific techniques: updates of neighboring value function approximations during the first iterations of the algorithm, and updates of value function approximations with the same climate spread cluster, but a different residual demand cluster.

The first approach is motivated by the idea that neighboring value function approximations, i.e., value function approximations with similar climate spread and residual demand clusters, are in general also similar. At least during the first iterations, the value function approximations of neighboring value functions are in most cases a better approximation of the real values than the initial values.

The second approach that we introduced to speed up convergence exploits some problem specific knowledge: if we consider two value function approximations that belong to the same climate spread cluster, but a different residual demand cluster, then the value function approximation belonging to the residual demand cluster with the higher residual demand is in general greater than or equal to the other value function approximation. If an update of a value function approximation violates this relationship, we use the updated value function approximation to update the value function approximations that violate this relationship.

In the third part of this chapter, we evaluated the quality of the solutions of our ADP algorithm. First, we tested the influence of different parameter choices of the algorithm. It turned out that the two problem specific approaches to update the value function approximations are crucial for the performance of the algorithm. Without these two methods, an algorithm with only one residual demand cluster performed better than an algorithm with multiple residual demand clusters. Applying both proposed update mechanisms, the algorithm with the additional residual demand clusters outclasses the approach with only one residual demand cluster.

Second, we tested the influence of the exploration method. Thereby, the modified ϵ-greedy method, a problem adjusted version of the ϵ-greedy method, performed slightly better than the standard ϵ-greedy approach. In contrast, the solutions obtained by Boltzmann exploration were of significantly lower quality. The tests we made have shown that a bad choice of a single parameter can result in an algorithm that does not work.

In the second part of the evaluation section, we compared the solutions of our ADP approach with those of a deterministic optimization based on expected fuel prices and with the solutions of our ISDP model. Both of our approaches perform better than the deterministic approach. The higher the uncertainty, the greater is the advantage of a stochastic approach. Compared to the ISDP model, the solution of the ADP model was slightly better. This can be explained by the ability of the ADP model to consider annual investments. As a consequence, the ADP model was able to find the optimal solution of a deterministic problem, whereby the ISDP model could not get this solution due to its restriction of allowing new investments only every four years.

While these results indicate that the ADP approach is superior to the ISDP approach, we do not think that this conclusion can be made. The ADP approach is an interesting new method to model the power generation expansion problem that can perform better than a standard SDP or our ISDP approach. However, the choice of the algorithm parameters is very problem specific, and a wrong choice of a parameter can result in an algorithm that performs very badly. Compared to the ADP approach, the ISDP method is less sensitive to the parameter choices.

In addition, the ADP approach is based on several assumptions that make an extension of the method difficult. For example, it is assumed that the value of the different types of new units is independent of each other. As is this not the case, we have introduced the different residual demand clusters. The problem is that these residual demand clusters are not suited to consider an additional investment alternative in our ADP model. Especially because of the high sensitivity of the ADP model concerning the parameter choices, we use the ISDP model for the case studies in the next chapter.

Chapter 5

Case Studies

In this chapter, we use our ISDP investment model to perform different case studies. The starting point for our analysis is a base scenario with fuel prices and carbon prices modeled as trended mean-reverting processes. We consider two investment alternatives: CCGT units and coal units. To obtain the contribution margins for the new units, we use our LDC model with adjustments.

In the first section of this chapter, we describe in detail the data we use for the case studies. In the following sections, we compare the solution of the base case with the solutions of alternative scenarios. With these alternative scenarios, we pursue different objectives. These objectives are to identify the influence of the stochastic process, the influence of other model-specific parameters like the discount rate, and the influence of political decisions like the nuclear phase-out on investment decisions.

In the second section, we examine the influence of the stochastic process and its parameters used to model fuel and carbon prices. As we have discussed in Section 1.4.2, there is no consensus in literature which process is pertinent. Two popular choices are the mean-reverting process and the GBM. As an alternative to our base scenario, we use a GBM for the simulation of fuel and carbon prices. To get an idea of the robustness of the solution of the base case, we also evaluate this solution with fuel and carbon prices simulated as GBM. These results are then compared with the evaluated profits of the best solution obtained using a GBM for the optimization. In our comparison, the optimal solution[1] of the base case performs only slightly worse than the one of the GBM.

---

[1]When we talk about the optimal solution in this chapter, we refer to the optimal solution of the approximated problem. This solution does not necessarily correspond to the optimal solution of the original problem.

Besides the stochastic process itself, also the parameters used for the process play an important role. Therefore, we examine how the optimal solution changes if different standard deviations or different drifts are used for the simulation of fuel and carbon prices. Especially the drifts used for the simulation have an important impact on new investments. The higher the drifts of fuel and carbon prices, the more profitable become new units.

In the third section of this chapter, we test the influence of other model parameters. We compare the solutions obtained using different discount rates and different depreciation methods. It is shown that a higher discount rate has a similar impact as an accelerated depreciation method: investments in new units become less profitable.

Besides the discount rate and the depreciation method, the method used to model investments by competitors has an important impact on the investment decisions of the considered operator. Our comparisons show that if investments by competitors are disregarded, the number of new units built by the considered investor decreases.

The investment in new units has also an influence on the profitability of the existing units. For that reason, the existing portfolio of an operator must be considered when investments in new units are optimized. We compare the investments of a new entrant in the market with those of existing generation companies. The new entrant in the market has the highest incentives to invest, the operator with the greatest portfolio has the lowest incentives.

In the forth section of this chapter we consider questions that are subject to political decisions. We start with an analysis of the consequences of a delayed or scrapped nuclear-phase out in Germany on investment decisions. It is shown that the longer existing nuclear units are allowed to operate, the lower are the incentives to invest new units.

Besides the nuclear phase-out, another issue with an important impact on investment decisions is the emission allowance allocation scheme. At the moment, emission allowances are allocated for free based on a fuel-specific benchmark. Alternative allocation schemes are a fuel-independent benchmark, or a full auctioning of emission allowances. Our analysis shows that the current scheme is the only one where coal units are preferred to CCGT units.

In the fifth section of this chapter, we examine another highly discussed question: the profitability of CCS units. In the before mentioned case studies we only allow investments in coal and CCGT units. With rising carbon prices, coal units get unprofitable. A possible solution to this problem is the development of the CCS technology. As the introduction of the CCS technology introduces additional costs and reduces the efficiency of CCS units, it is uncertain whether this technology will become competitive. The competitiveness of CCS units depends mainly on future carbon prices. Due to the difficulties of simulating carbon prices appropriately, we use our LP-based fundamental electricity market model to examine this question, as it allows the determination of carbon prices based on fundamentals. In the considered case, CCS units are the preferred choice on average.

5.1 Input Data

In this section, we describe the input data that we use for our base scenario. The optimization horizon for this scenario is 24 years. Basically, we use the same data that we already used for the evaluation of the different fundamental electricity market models. This data was described in Section 2.4.1. As the focus on the evaluation of the fundamental electricity market models was on past periods, we did not need to specify assumptions for future years. In the following, we focus on the description of this future data.

We first describe our assumptions about future electricity demand, before we specify our assumptions about the development of electricity supply. Next, we present the two investment alternatives, before we conclude this section with a description of the stochastic process and its parameters used to model fuel and carbon prices.

5.1.1 Electricity Demand

Recent studies in the electricity sector describe different scenarios for the development of electricity demand. There are scenarios assuming a rising demand (e.g., ATK [ATK09], the 2%-scenario in EWI and Prognos [EP07]), others expect a constant demand (e.g., the constant scenario in DENA [DEN08]), while other scenarios are based on a decreasing demand (e.g., CONSENTEC et al. [CON+08], the other scenarios in [EP07]). For our case studies, we assume a constant demand.

5.1.2 Electricity Supply

Basis for the case studies is our power plant database described in Section 2.4.1. Besides existing power plants, we also consider power plants under construction,

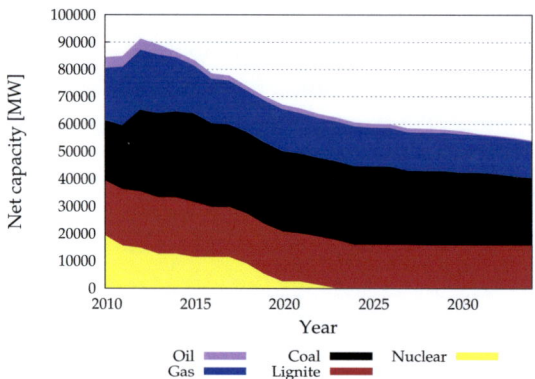

Figure 5.1: Development of available capacity.

	2010	2015	2020	2030	2040	2050
Offshore wind	0.4	3.6	10.0	23.0	32.8	37.0
Onshore wind	25.3	27.2	28.1	29.8	32.5	34.0

Table 5.1: Expected installed wind power capacity in GW. Source: Nitsch [Nit08].[2]

or power plants that are planned and have a high probability that they are realized. Concerning the nuclear phase-out, we assume that the new government is going to extend the operational life of nuclear units by ten years.

An overview of the available capacity of conventional power plants is given in Figure 5.1. We assume that the operational life of CCGT units and oil units is 40 years, those of coal and lignite units 45 years, and the one of gas turbine units 50 years [DEN08]. To determine the investments of competitors within the optimization horizon, we use the investment module based on the shortage of supply described in Section 3.3.1.

The expected amount of installed wind power capacity is shown in Table 5.1. To obtain the hourly wind production, we use the hourly time series for on-shore and off-shore wind described in Section 2.4.1, adjusted by the corresponding installed capacity. For run-of-river plants, pumped storage and hydro reservoirs, we assume that the available capacity remains constant.

---

[2]For the years 2030 and 2040, only the sum of the installed wind power capacity is given in tabular form. The breakdown to onshore/offshore has been derived from a figure. Therefore, minor imprecisions are possible for these years.

	Coal unit	CCGT unit
Investment costs (million €)	1,000	500
Capacity (MW)	1,000	1,000
Efficiency	0.46	0.60
Planning & construction time (years)	4	4
Annual fixed costs (€/MW)	20,000	15,000
Economic useful life (years)	25	20

Table 5.2: Input data used for the investment model. Sources: DEWI et al. [DEW+05], own assumptions.

5.1.3 Investment Alternatives

Except for the case study analyzing the profitability of CCS units in Section 5.5, we consider two investment alternatives: CCGT units and coal units. We allow the construction of up to ten new units of each type. The parameters of these units are given in Table 5.2. The other technical parameters are assumed to be the same as the ones of units of the corresponding technology described in Section 2.4.1.

We use a straight-line depreciation to calculate the part of the investment costs that is assigned to the years of operation within the considered optimization horizon. We select the straight-line depreciation, as this is the method used by the big four generation companies in their annual reports [EnB08; EON08; RWE08; Vat08]. Besides the depreciation method, we need to specify the economical useful life of new units. EnBW and Vattenfall apply an economic useful life between 15 and 50 years for power stations, RWE assumes 15 to 20 years of useful life. E.ON uses an economic useful life of 10 to 65 years for technical equipment, plants and machinery.

For our case studies, we assume an economic useful life of 20 years for CCGT units, and 25 years for coal units. The longer useful life of coal units is in line with current literature (see, e.g., Konstantin [Kon08], OECD and IEA [OI05]).

To discount future cash-flows, we use an interest rate of 7%. This value is in the interval used by other publications (e.g., 5% are used in [OI05], 7% by Weber [Web05], and 7.5% in [Kon08]).

5.1.4 Fuel and Carbon Prices

Concerning the fuel and carbon price development, forecasts are even more diverging than those for electricity demand. We gave an overview of different car-

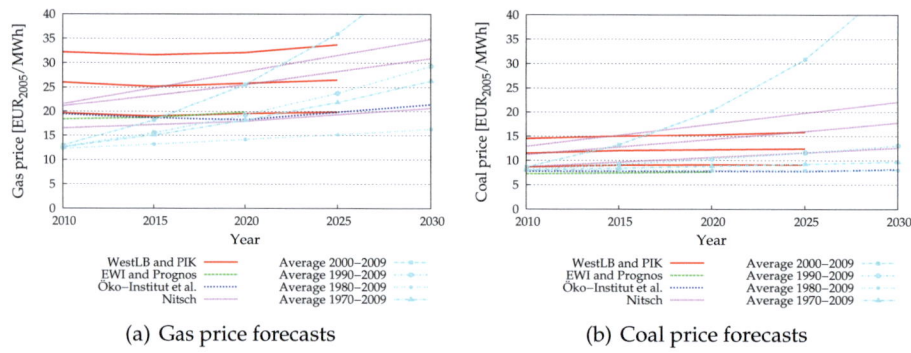

(a) Gas price forecasts   (b) Coal price forecasts

Figure 5.2: Fuel price forecasts of recent studies in Germany.

bon price forecasts in Table 1.2 on page 21. Those forecasts predict rising carbon prices. In 2020, carbon prices between 30 €/t and 40 €/t are expected. However, it should be stated that there are also other opinions. Without specifying exact forecasts, ATK [ATK09] assumes that under current conditions, lower carbon prices are realistic for longer periods.

Following a survey amongst market participants (Point Carbon [Poi09]), about 29% expect carbon prices in 2020 between 35 €/t and 50 €/t. About 21% predict prices between 25 €/t and 35 €/t, about 10% assume that prices will be higher than 50 €/t, while the rest of the market participants expects prices lower than 25 €/t. For our model, we assume carbon prices of 15 €/t for 2010. For the following years, we assume an annual increase by 7%, resulting in expected carbon prices of 29,50 €/t in 2020. We have chosen these relatively low prices, as our model is based on real prices.

In 2020, phase III of the EU ETS ends. For the years afterwards, the legal framework for emissions trading is not yet fixed. Given the ambitious objective of reducing the emissions by 80–95% in 2050 compared to 1990 [BMU09], we assume that carbon prices continue to increase by 7% per year.

Figure 5.2 depicts gas and coal price forecasts of four recent studies. The prices are real prices with base year 2005. Nitsch [Nit07] and WestLB and PIK [WP09] use three different price scenarios each (low, middle, high), EWI and Prognos [EP07] and Öko-Institut et al. [ÖI+08] use one scenario. In addition to these scenarios, we depicted four alternative scenarios. These four scenarios are based on average EEX spot prices for January to October 2009 used as a base price and different constant annual growth rates. These annual growth rates are estimated from his-

	Initial price	Average annual price change [%]	StDev	Mean reversion rate
Gas	18 €/MWh	1.4	0.15	0.14
Coal	9 €/MWh	1.0	0.1	0.14
CO_2	15 €/t	7.0	0.2	0.14

Table 5.3: Values used for the fuel and carbon price simulation.

torical data. To obtain the annual growth rates, we took the average annual import prices [Des09a] and deflated them by the consumer price index [Des09c]. Based on different periods, we determined average annual growth rates, which were used to derive the curves denoted with *Average* in Figure 5.2.[3]

Depending on the considered period, annual growth rates vary significantly. Annual growth rates for the period between 2000–2009 are much higher than the ones expected by the considered surveys. For gas prices, the growth rate of the years 1980–2009 corresponds best to the expected values of the surveys. For coal prices, the growth rate of the period from 1970 to 2009 comes closest to the expectations.

A consequence of the current economic crises can be seen in Figure 5.2: the expected fuel prices for 2010 obtained by our calculations based on the fuel prices of 2009 are—especially for the gas price—significantly below the forecasts of the studies. As we assume that fuel prices will return to a higher level when the economy recovers, we use higher prices as basis for the calculation of the mean-reverting level for our simulations.

For the case studies in this chapter, we simulate the logs of fuel and carbon prices as a trended mean-reversion process. Such a process is used for fuel prices for example by Pindyck [Pin99], for carbon prices by Spangardt and Meyer [SM05]. An important parameter of a mean-reverting process is the rate of mean-reversion. Pindyck states that the rate of mean-reversion is slow, but he does not give exact numbers. For oil prices, he estimates a half-life of the mean-reverting process of about five years. Based on the half-life h, we can determine the rate of mean-reversion as $\ln(2)/h = 0.14$. For our case studies, we use this value. Table 5.3 summarizes the values we use for the fuel and carbon price simulations.

As discussed in Section 1.4.3, modeling the long-term development of carbon prices is difficult. For that reason, many long-term electricity market models either neglect carbon price uncertainty or use some predefined scenarios. While there is

---

[3]Data for the year 2009 is based on the period from January to September 2009.

	Gas	Coal	CO_2
Gas	1	0.55	0.5
Coal		1	-0.5
CO_2			1

Table 5.4: Correlations between logs of price changes.

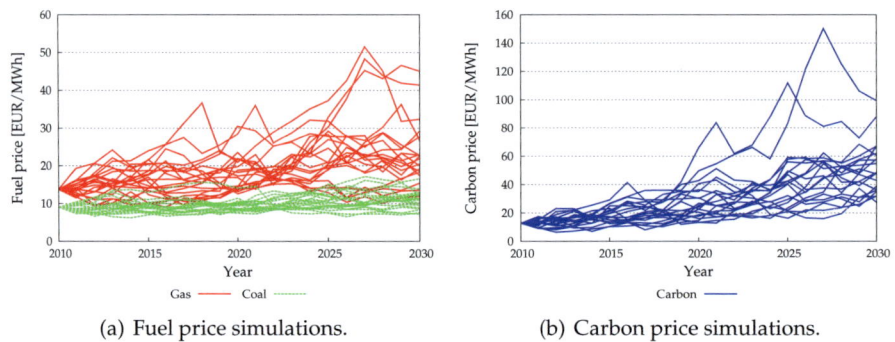

(a) Fuel price simulations.   (b) Carbon price simulations.

Figure 5.3: Example for 20 fuel price and carbon price simulations.

no evidence that carbon prices are mean-reverting, we think that modeling carbon prices as mean-reversion process captures the uncertainty related to the price development better than deterministic prices or a few different scenarios.

We assume that the random increments of the different mean-reversion processes are correlated. For this purpose, we use the correlation values specified in Table 5.4. It should be noted that these are the correlations of the logs of price changes, and not the correlations between the prices itself.

Figure 5.3 shows an exemplary outcome of 20 fuel price and carbon price simulations with the described settings. It can be seen that the uncertainty about future gas and carbon prices is higher than the uncertainty about future coal prices. While all three price processes follow a mean-reversion process, the low mean-reversion rate allows remarkable deviations from the expected mean.

5.2   Influence of the Stochastic Process

In Section 1.4.2, we discussed the difficulties related to fuel price modeling. Basically, there are two issues: the stochastic process used for the simulation, and the parameters of the chosen process. For both points, there is no consensus about the right choice. In the following, we first compare the solutions based on fuel prices simulated as a trended mean-reverting process with those based on a GBM. Next, we examine the impact of wrong choices concerning the parameters used for the fuel and carbon price simulation.

5.2.1 Alternative Stochastic Process

Two popular choices for the stochastic process used to model fuel prices are the correlated GBM and a (trended) correlated mean-reverting process. For the base scenario in this chapter, we have chosen a trended mean-reverting process. In this subsection, we examine the influence of a possibly wrong choice of the stochastic process.

For this purpose, we compare solutions obtained with the trended mean-reverting process of our base scenario with the solutions of two alternatives: a correlated GBM, and a trended mean-reverting process with mean-reversion rates of 0.7. For both alternatives, we use the same expected prices, standard deviations and correlations for the fuel price simulation.

Besides a comparison of the solutions, we also analyze how the solution of the base case performs, if the evaluation is performed using a different stochastic process for the price simulations. To get an idea of the consequences of an inaccurate choice of the stochastic process, we compare the evaluated value of the base case under the alternative stochastic process with the evaluated value of the best solution obtained when the corresponding stochastic process is also used for the optimization.

Comparing the optimal solutions of the different stochastic processes delivers a result that may be surprising at first glance. On average, more new units are built when a GBM is used compared to both mean-reverting processes. The least amount of new capacity is added when we use mean-reversion rates of 0.7. Figure 5.4(a) visualizes these results.

It is remarkable that the incentives for investments in new units are low: the first new units are built in 2022. Almost exclusively new CCGT units are built. These results are in line with results reported by WestLB and PIK [WP09], but may seem a bit surprising given the current discussion whether CCGT units are a profitable investment alternative to coal units. As we will see later in this chapter, the low degree of investments and the preference of CCGT units are caused by the assumption of a delayed nuclear phase-out and the full auctioning of emission allowances.

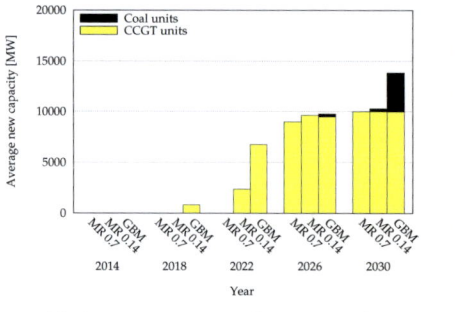

(a) Average amount of new capacity.

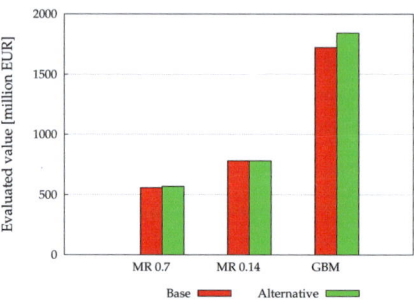

(b) Evaluated values.

Figure 5.4: (a) Average amount of new capacity for different scenarios. (b) Value of the optimal solution of the base case evaluated using different stochastic processes. The right bars are the values of the best solution found using the corresponding stochastic process.

If a GBM is used instead of a mean-reverting process, more units are built and units are built earlier. The reason for the higher average amount of new capacity and the earlier construction of this capacity of the solution based on a GBM is the following: Under the expected fuel prices, investments are relatively unattractive. However, if fuel prices develop such that one technology gets more attractive compared to the other (in our scenarios the more attractive technology are almost always CCGT units, i.e., coal prices rise stronger than gas prices), investments in one of the two technologies may become profitable. If a mean-reverting process is used to simulate fuel prices, this profitability may only be a temporary one as prices revert to the expected prices. When fuel prices are assumed to follow a GBM, this is not the case. Hence, it can be expected that the profitability for the corresponding technology persists. The same fact explains why a higher mean-reversion rate reduces the amount of new units on average.

Figure 5.4(b) shows the average discounted value of the best solution of the base case evaluated with the different stochastic processes. For a better comparison, the figure also shows the average discounted value of the best solution obtained when the corresponding stochastic process is used during the optimization. Two conclusions can be drawn from the data depicted in this figure. First, the higher the uncertainty is (i.e., the lower the mean-reversion rates are), the higher are the expected profits. However, not only the expected profits increase with uncertainty, but also the risk. The standard deviation of the evaluated returns is over ten times higher for the GBM compared to the mean-reverting process with mean-reversion

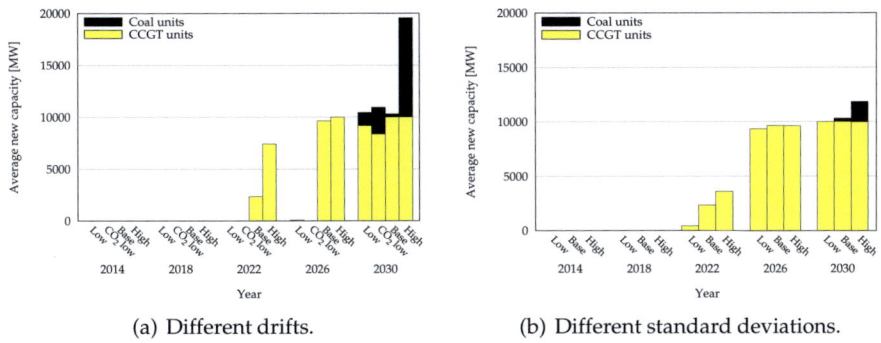

(a) Different drifts.

(b) Different standard deviations.

Figure 5.5: Average amount of new capacity for different scenarios.

rates of 0.7. Secondly, the best solution of the base case performs relatively well if another fuel price process is used for the evaluation. Compared to the best solution of the GBM, the base case solution is about 6.5% worse. In case of the mean-reverting process with higher mean-reversion rate, the difference is below 2%.

5.2.2 Alternative Parameters

Besides the choice of the stochastic process used for the fuel and carbon price modeling, the parameters chosen for this process play an important role. For the trended mean-reverting process, there are three important parameters: the mean-reversion rate (which we analyzed in the previous subsection), the drift of the process and the standard deviation. In this subsection, we examine the impact of the drifts and the standard deviations.

Let us start with the drifts. We consider four different scenarios: the base scenario, a scenario with higher drifts, a scenario with lower drifts, and a scenario where only the drift of carbon prices is reduced. The base scenario has an expected annual increase of gas prices of 1.4%, an increase of coal prices of 1.0%, and an increase of carbon prices of 7.0%. For the scenario with the higher drifts, we assume that the drifts of all three processes are increased by 50%, while for the scenario with lower drifts, all drifts are reduced by 50%. In the last scenario, only the drift of carbon prices is reduced by 50%.

Figure 5.5(a) shows the average newly installed capacity for these four scenarios. If all drifts are simultaneously changed, then the higher the drifts, the more units are built. Investments are more attractive, because electricity prices tend to rise

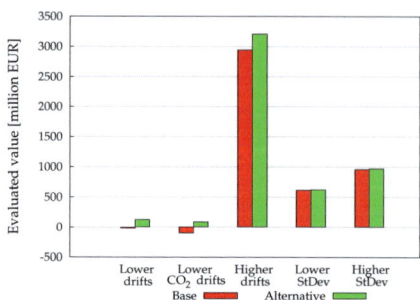

Figure 5.6: Value of the optimal solution of the base case evaluated using different parameters for the stochastic process. The right bars are the values of the best solution found using the corresponding parameters also during the optimization.

more than the variable costs of new units. The reason for this behavior is that electricity prices are often set by less efficient units, for which the costs rise more than the costs of new units. If only the carbon price drift is reduced, coal units get slightly more competitive compared to the other scenarios. While in this scenario, the lowest amount of new CCGT units is built, CCGT units are still the preferred choice on average.

Figure 5.5(b) depicts the results for different scenarios concerning the standard deviations used for the simulation of the fuel prices and the carbon prices. For the *low* scenario, we use only half of the standard deviations of the base scenario. For the *high* scenario, we increase the standard deviations by 50%. It can be seen that the higher the standard deviations, the more units are built. The reason for this behavior is that higher standard deviations lead to more extreme price differences. During times of extreme price differences, huge profits can be obtained by the cheaper technology. Even the unfavorable technology may profit from the high prices. The explanation is the same as the one for the higher degree of investments when higher drifts are used: if electricity prices are set by units of the unfavorable technolgy, electricity prices rise more than the costs of the new units.

Figure 5.6 illustrates how the best solution of the base case performs if other values for the standard deviations or the drifts are used for the price simulation during the evaluation. It can be seen that the drifts of the fuel and carbon prices have a significant impact on the profitability of new investments. A higher drift increases the profitability of new investments in our examples. While the best solution of the base case performs satisfactorily if drifts are higher than assumed, it results in negative evaluated values if drifts are lower. In the base case, the first new units are built in 2022. If drifts are lower, almost no new units are built before 2030. The losses obtained with the solution of the base case evaluated with lower drifts are due to a too early construction of new units. However, it can be assumed that the performance of the base case solution is in reality not as bad as the results shown in Figure 5.6: The construction of new units is not started before 2018. It seems plausible that if drifts deviate strongly from the expected drifts, then this should be realized until 2018. A re-optimization with updated values will then deliver a better solution.

Compared to the drifts, the standard deviations used for the simulation of the fuel and carbon prices play a less important role. Changing the standard deviations has a similar, less distinct effect as changing the mean-reversion rates: It increases the uncertainty, and with it the evaluated value of the best solution as well as the risk. For the considered examples, the best solution of the base case performs only slightly more than 1% worse than the best solution found when the corresponding alternative standard deviations are used for the optimization.

5.3   Influence of Other Model-Specific Parameters

Besides the process used for the fuel and carbon price simulation, there are several other parameters that have an impact on the results of our model. These parameters are the discount rate, the depreciation method, the method used to determine investments of competitors, and the consideration of an existing portfolio. In this section, we analyze the influence of these parameters.

5.3.1   Depreciation Method and Discount Rate

Two parameters that can be used to consider different degrees of risk aversion are the discount rate and the depreciation method. A higher discount rate requires a faster amortization or higher returns for an investment to be carried out. Hence, a risk averse investor may choose a higher discount rate than a risk-neutral investor. A similar effect can be achieved with the selection of depreciation method. A straight-line depreciation assumes that new units realize constant profits during their useful life. Accelerated depreciation methods are based on the idea that the depreciation is higher during the first years of operation. In our model, an accelerated depreciation method leads to similar results as a higher discount rate. This can be seen in Figure 5.7.

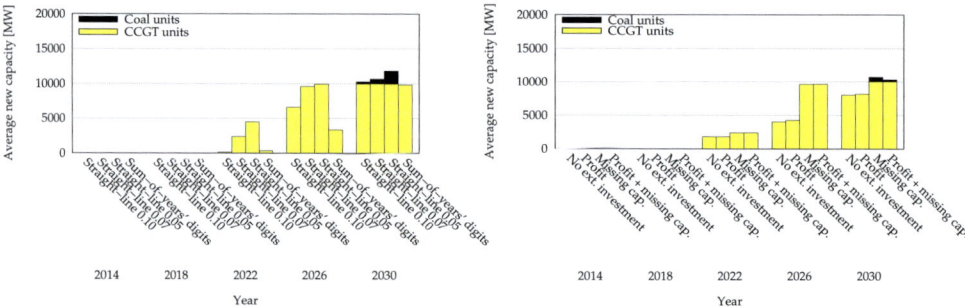

Figure 5.7: Influence of the discount rate and the depreciation method.

Figure 5.8: Influence of the method used to determine investments by competitors.

In this figure, the solution based on a straight-line depreciation and a discount rate of 5%, 7% and 10% are compared with the results of a sum-of-years' digits method with a discount rate of 7%. The lower the discount rate, the earlier are new units built. If the sum-of-years' digits method is applied, fewer units are built than with a straight-line depreciation. Under different conditions compared to the base scenario, e.g., different fuel prices, the choice of a higher discount rate or an accelerated depreciation method may also change the relative profitability of CCGT units compared to coal units. Thereby, a higher discount rate or an accelerated depreciation method favours less capital intensive technologies, in our case CCGT units.

5.3.2 External Investments

In our model, we optimize the investments of a single operator. As the profitability of new units depends on the existing power generation fleet, we have to consider the decisions of other generation companies. In Section 3.3, we presented two methods how investments of competitors can be considered in our model. The impact of these methods is examined in this subsection.

Figure 5.8 shows the results for different methods incorporating external investments. The method based on the shortage of supply is denoted with *Missing cap.*, the method considering the expected profitability with *Profit*. A combination of both methods is denoted with *Profit + missing cap.*. Besides these three scenarios, we also investigate the case when no external investments are considered (denoted as *No ext. investments*). For the investments based on the expected profitability, we use threshold values of 15 €/MWh ($CDS^{\text{Threshold}}$) and 10 €/MWh ($CSS^{\text{Threshold}}$). The price effect of a new unit (p^{Unit}) is assumed to be 2 €/MWh.

The results depicted in Figure 5.8 show that the external investments based on the expected profitability of new units have only little impact. The reason is that for the determination of the expected profit, we only use past data. This myopic approach does not consider that old units are decommitted in the future and need to be replaced. Without the decommitment of old units, new units are rarely profitable in our test cases.

Comparing the results of the method based on the shortage of supply with the method disregarding investments by competitors, the importance of the consideration of other investments can be seen. Disregarding investments of competitors results in a significantly lower degree of investments. By limiting the amount of available capacity, operators can earn the VoLL, which is (with 2,000 €/MWh in our model) much higher than the prices obtained when demand can be satisfied.

5.3.3   Consideration of an Existing Portfolio

The construction of new units influences electricity prices and the electricity production of other units. Hence, it also changes the contribution margin of other units. As a consequence, an investment model should take into account the existing portfolio of the considered operator.

In this subsection, we compare the optimal solution of a new entrant to the market, with those of an operator with an existing portfolio. For this comparison, we choose exemplarily two operators, EnBW and RWE. When we model external investments with the investment module based on the shortage of supply, then the optimal decisions are very similar in all three cases. The reason is the following: Under most fuel price scenarios, the new capacity added by the optimal decision—even for a new entrant in the market—is lower than the capacity required to satisfy the peak load. Hence, independent of the decision of the considered operator, additional capacity is built. The total amount of new capacity is almost independent of the decision of the considered operator. In such a case, the investment decisions are independent of the existing portfolio, because external investments are triggered if the considered operator does not invest. Note that this independence is only given in this special case.

To highlight the importance of the existing portfolio, we also consider an extreme scenario: no external investments. In this case, the optimal decision depends strongly on the existing portfolio. Figure 5.9 depicts the optimal solution of a new

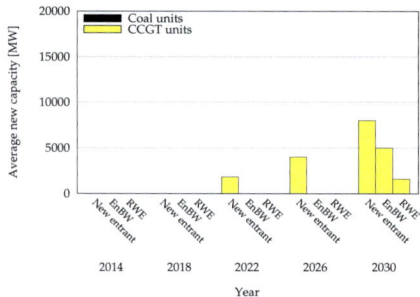

Figure 5.9: Average new capacity if no external investments are considered.

entrant in the market, EnBW and RWE. In all cases, only CCGT units are built. The higher the existing capacity of the considered operator, the lower the number of new units that are built on average. The reason for this behavior is that the construction of new units leads to a lower electricity price. Note that in this example, the ignorance of external investments results in a shortage of supply such that the VoLL sets the electricity price during several hours each year.

5.4   Influence of Political Decisions

Despite the deregulation of the electricity sector, the profitability of new investments is still strongly influenced by political decisions. In Germany, there are two important political decisions at the moment: the nuclear phase-out and the emission allowance allocation scheme. Both issues are not finally decided, therefore we analyze the impact of different possible scenarios. We start with the nuclear phase-out, and consider afterwards the emission allowance allocation scheme.

5.4.1   The Nuclear Phase-Out

A controversial debate exists in Germany about the nuclear phase-out. While the majority of the population is in favor of the nuclear phase-out [Gre09], the new government plans to delay or scrap the phase-out [CF09]. A detailed analysis of all the implications of the nuclear phase-out on the electricity market is beyond the scope of this thesis, therefore, we focus on the effect on investment decisions. For this purpose, we compare the optimal investment decisions of a new entrant into the market under three different scenarios. The first scenario is based on the assumption that the nuclear phase-out is realized as planned. The second scenario assumes an extension of the allowed time of operation of ten years for every nuclear plant. In the third scenario, the time of operation of nuclear plants is not restricted.

Figure 5.10 shows the average amount of new capacity built by the considered investor for these three scenarios. The optimal solution consists in all three cases mainly of new CCGT units. Only in case of a normal nuclear phase-out, a greater amount of coal units is built. However, these coal units are only built because a great amount of new capacity is required and the number of new units of each type is restricted to ten in these case studies.

The planned nuclear phase-out reduces the available capacity in the next years. This missing capacity makes investments in new units more attractive. If the nuclear phase-out is postponed or scrapped, then the construction of new units is also postponed. In the public debate it is often argued that the profiteers of a delayed or scrapped nuclear phase-out are the generation companies. However, this is only partly true. Only those generation companies with nuclear units benefit from a modification of the nuclear phase-out. New entrants into the market are losers: While the expected discounted value of the optimal solution with the normal nuclear phase-out is $1,334$ million euro, a delayed phase-out results in expected profits of 788 million euro. If the nuclear phase-out is scrapped, the expected profits drop to 413 million euro.

5.4.2 The Emission Allowance Allocation Scheme

During the phase I and phase II of the EU ETS, new units have received the emission allowances for free. The amount of free emissions a new plant receives currently in Germany depends on the fuel of the plant. The best available technology with the same fuel is used as benchmark. This benchmark is for CCGT units 365 g CO_2/kWh, for coal units 750 g CO_2/kWh. The amount of free emission allowances is based on this benchmark and an assumed number of 7,500 full load hours.

The different benchmarks for CCGT units and coal units are criticized by environmental organizations (e.g., BUND [BUN06]). Instead of a fuel-specific benchmark, a fuel-independent benchmark, or—even better—an auctioning of the emission allowances is claimed. Even Statkraft, a generation company, supports these claims [Sta06].

During phase II of the ETS, the EU prescribed that at least 90% of the emission allowances are allocated free of charge. For phase III, which starts in 2013, the

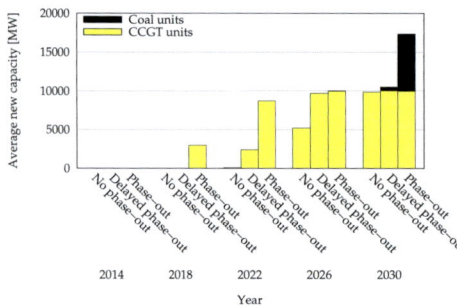

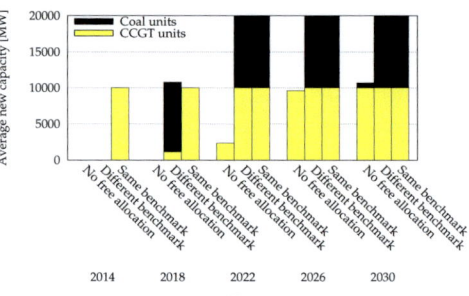

Figure 5.10: Incluence of the nuclear phase-out.

Figure 5.11: Influence of the emission allowance allocation scheme.

EU states that *auctioning should [...] be the basic principle for allocation, as it is the simplest, and generally considered to be the most economically efficient, system* (Directive 2009/29/EC amending Directive 2003/87/EC[4]).

In the following, three different emission allowance allocation schemes are compared: the current system based on a fuel-specific benchmark, an alternative system using the same benchmark for CCGT and coal units and a full auctioning of emission allowances. We refer to the alternative system with only one benchmark as *fuel-independent benchmark*. For the alternative system, we use the value of 557.5 g CO_2/kWh as benchmark, corresponding to the average of the current benchmarks.

Figure 5.11 depicts the average amount of newly installed capacity under these three allocation schemes. Two different effects can be observed in this figure. First, a free allocation of emission allowances increases the incentive to invest in new units. If emission allowances are fully auctioned, only 2,400 MW of new capacity are built on average until 2022. If emission allowances are allocated for free, the construction of new units starts earlier, and more new units are built. Comparing the two schemes with free allocation of emission allowances, it can be seen that the scheme with a fuel-independent benchmark yields immediate investments. The fuel-specific benchmark attracts new investments four years later. The reason is that CCGT units are less capital intensive. Less capital intensive units exhibit a lower risk, and are therefore built earlier than more capital intensive units.

The second remarkable result is that the current allocation scheme is the only one in which coal units are preferred to CCGT units. Based on the current scheme,

---

[4]http://eur-lex.europa.eu/LexUriServ/LexUriServ.do?uri=OJ:L:2009:140:0063:0087:EN:PDF, accessed on December 02, 2009.

9,200 MW of new capacity provided by coal units and 1,200 MW of new capacity provided by CCGT units is built on average up to 2018. Under both alternative allocation schemes, coal units are almost exclusively built when the maximal number of CCGT units is reached.

5.5 Evaluation of the Profitability of CCS Units

The CCS technology might play an important role to reach global climate goals and reduce CO_2 emissions. Today, only a pilot capture project exists in Germany, Vattenfall's Schwarze Pumpe. Nevertheless, it is assumed that the CCS technology gains in importance in the future. However, there are still some open questions. Besides technological questions, an open issue is the question of the competitiveness of CCS units.

On the one hand, the reduced emission of CO_2 results in savings compared to non-CCS units. On the other hand, the capture process reduces the efficiency of the CCS plant. In addition, the CCS equipment causes additional costs. The profitability of CCS units depends greatly on future carbon prices. McKinsey [McK08] assumes that a carbon price between 30–48€ per tonne CO_2 is required for CCS units to be competitive. Many forecasts are within this region. However, as we have discussed in Section 1.4.3, carbon price forecasts are subject to significant uncertainty.

We outlined in Section 2.3.5 how carbon prices can be determined within a fundamental electricity market model based on an LP formulation. In this section, we apply our LP-based fundamental electricity market model to examine the profitability of coal-fired CCS units.

The advantage of a carbon price determination based on an electricity market model is that it explicitly considers the interdependence of fundamental factors like fuel prices and the existing power generation fleet (and hence the demand of emission allowances). Compared to other carbon price forecasts, this approach has two shortcomings: First, the carbon price determination of different years is completely independent, it does not consider the possibility of banking. Secondly, as we model only the German electricity market, we consider only a part of the emission allowances market. To account for these shortcomings, we use the average of the price determined by our model and the predicted price. In addition, we add a price cap corresponding to twice of the predicted price.

Investment costs (million €)	1,500
Capacity (MW)	1,000
Efficiency	0.37
Carbon capture rate	0.9
Planning & construction time (years)	4
Annual fixed costs (€/MW)	25,000
Economic useful life (years)	25

Table 5.5: Parameters of the CCS units used for the case study. Sources: McKinsey [McK08], own assumptions.

To be able to determine the carbon price with our model, we need to specify the amount of allowed emissions per year. This amount is so far only fixed until 2012. The EU formulates in Directive 2009/29/EC amending Directive 2003/87/EC the planned general framework for the post 2012 EU ETS. Following this directive, we assume that for the years after 2012 the amount of allowed emissions is linearly reduced every year by 1.74% of the average emissions in phase II of the EU ETS.

Table 5.5 depicts the parameters of the CCS unit used for the case study. Compared to non-CCS units, it is assumed that investment costs are increased by 50%, while efficiency is reduced by 9%. Furthermore, it is assumed that 90% of the emitted CO_2 can be captured [McK08]. Concerning the construction time and the economic useful life, we use the same values as for non-CCS coal units, while we assume that annual fixed costs are increased to 25,000 €/MW. Due to the significantly longer running times of the LP-based electricitiy market model compared to the LDC model, we use an optimization horizon of 20 years. We allow a maximum of eight new units of each type (CCGT, coal, coal-fired CCS). We use the weighted, deterministic LP model for the optimization of the electricity production. We opt for the deterministic model to reduce the running times.

Figure 5.12(a) depicts the average amount of new capacity. Similar to the case studies carried out before, non-CCS coal units are not competitive compared to the alternatives. While the choice of the technology depends on the fuel price development, it can be seen that on average CCS units are the preferred choice. Compared to the case studies based on the LDC electricity market model, fewer new units are built.

A reason for this behavior can be seen by looking at Figure 5.12(b). This figure shows the average carbon prices based on the simulation used for the case studies in the previous sections of this chapter, and the average carbon price observed when the optimal solution of the LP model is applied. Up to the year 2022, we

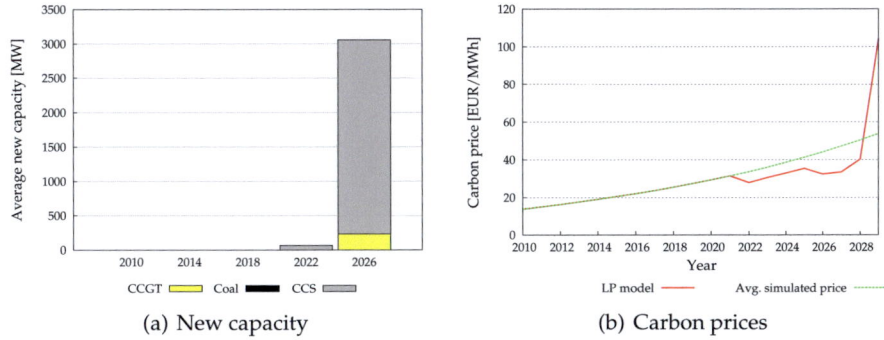

(a) New capacity (b) Carbon prices

Figure 5.12: Average results of the case study evaluating the profitability of CCS units.

do not have separate values for the solution of the LP model. To save time, we do not run the LP-based electricity market model if no new units are built, as the contribution margins for a new entrant to the market are zero in this case. As a consequence, we do only determine the carbon price with our LP model if new units are built. In Figure 5.12(b), this is the case from 2022 until 2029. Except for the year 2029, the average prices obtained with the LP model are below average simulated prices. As we have discussed before, a higher price level increases the profitability of new units. Due to the lower price level, fewer units are built in this case study.

Two interesting effects concerning the price determination with a fundamental model can be noticed in Figure 5.12(b). First, carbon prices rise every year after 2022 except for the year 2026. The increase of the prices is caused by a reduction of the available amount of allowed emissions. The decrease of the carbon price in 2026 is due to the construction of new units. On average, three new units are added in 2026. These efficient new units replace older less efficient units, and hence decrease the amount of required emission allowances. Note that this effect reduces the incentives for the investment in new units: compared to the LDC model with simulated carbon prices, the construction of new units in the LP model has a greater impact on electricity prices, as it also decreases carbon prices.

The second remarkable thing is the extreme price increase in 2029. In 2029, the amount of allowed emissions is not sufficient for the production of the requested electricity. As a consequence, the constraint restricting the amount of emissions results in unsatisfied electricity demand. During times of unsatisfied demand, the expensive VoLL sets the electricity price. As a consequence, the marginal value for additional emission allowances gets extremely high. Due to the price cap we introduced, the carbon price is set to 108 €/t CO_2, corresponding to the double of the expected carbon price in 2029. However, if an additional CCS unit is added in 2029, average carbon prices drop to 71 €/t CO_2. In reality, the construction of one single unit does not have such huge effects on the carbon price.

One possibility to circumvent this effect would be to allow a transfer of unused emission allowances from one year to the next. However, this would introduce a path-dependence that is not compatible with our dynamic programming formulation of the investment model. Despite this undesirable sensitive behavior of the price determination, the results of the case study are interesting, as they indicate that carbon prices will rise to a level making the CCS technology profitable.

5.6 Summary

In this chapter, we applied our ISDP investment model to different case studies. We defined a base case and compared it to different alternative scenarios varying some parameters of the model. In the base scenario, we used a trended mean-reverting process for the fuel and carbon price simulation. We assumed that emission allowances are fully auctioned, and that the operational life of nuclear units is extended by ten years. The considered investment alternatives were coal units and CCGT units. In the base scenario, we optimized the power plant investments for a new entrant to the market. The optimization horizon was 24 years. To determine the contribution margin of new units, we used our LDC electricity market model.

In the base scenario, investment incentives are low. The construction of new units is not started before 2018. Almost only CCGT units are built. Similar results, namely a low degree of investments and uncompetitive coal units, have been recently reported by WestLB and PIK [WP09]. These results raise the question whether the current market design offers sufficient incentives for new investments, an issue controversially discussed in literature (see, e.g., Oren [Ore05] and Joskow [Jos08]).

One of the main issues related to power plant investments is the fuel and carbon price uncertainty. While the methods proposed in this thesis allow to consider many different fuel and carbon price scenarios, the results still depend on the chosen stochastic process and its parameters used for the simulation. As there is no consensus neither about the pertinent process nor about the exact parameters, we have compared the base scenario with scenarios using a different stochastic process or different parameters for the trended mean-reverting process. Thereby, we examined two questions: First, we studied how the best solution changes depending on the chosen stochastic process and its parameters. Secondly, we analyzed how the solution of the base case performs, if a different stochastic process or different parameters are used for the evaluation. The latter test gives us an idea about the consequences of an inaccurate choice of the stochastic process.

Besides the mean-reversion process, the GBM is a popular choice for modeling fuel prices. For this reason, we compared the solutions of the trended mean-reversion process used in the base case with the solutions obtained with a GBM. In addition, we also used a trended mean-reversion process with higher rates of mean-reversion. Due to the mean-reverting property, the long term uncertainty of a mean-reverting process is smaller than the one of the GBM.

The results of our comparison may be surprising at first sight: the higher the uncertainty, the more new units are built. In the same time, the construction of new units is started earlier. The reason for these results is the following: the expected fuel prices offer only low incentives for investments in new units. However, if coal prices or carbon prices rise stronger than expected, then new CCGT units become profitable. If fuel prices follow a GBM or a mean-reverting process with a low rate of mean-reversion, it can be assumed that these favorable fuel prices persist at least for some years. This period may be long enough to make new CCGT units profitable. If fuel prices follow a mean-reverting process with high mean-reversion rates, then fuel prices return quickly to the unattractive expected price level.

The higher the uncertainty, the higher is the expected value of the best solution found by our model. In the same time, the standard deviation of the evaluated returns does also increase. If the best solution of the base case is evaluated based on the GBM or based on a mean-reversion process with a higher rate of mean-reversion, it performs only slightly worse than the best solution found by an optimization based on the alternative stochastic process. Hence, even if we chose the wrong stochastic process, the solution we obtain is of good quality.

Besides the type of the stochastic process and the rate of mean-reversion, we also tested the influence of the drift and the standard deviation used for the fuel and carbon price simulation. Our tests showed that especially the drifts have an important impact on the solution. The higher the drift, the more profitable are new units. As new units are in general more efficient than older units, fuel and carbon costs of new units increase below-average. If less efficient units set the electricity price, the profit of the more efficient new units increases with rising fuel prices. As a consequence, the higher the drifts, the more new units are built and the higher are the expected profits. If we evaluate the solution of the base case with an alternative mean-reverting process with drifts increased by 50%, then the solution of the base case performs 8% worse than the one of the alternative. However, if we use an alternative process with drifts reduced by 50%, i.e., the base scenario overestimates the drifts of the fuel price processes, then the solution of the base scenario produces losses on average. As future fuel prices are overestimated, the profitability of new units is overestimated, and new units are built under unfavorable conditions. However, as in both cases the construction of new units is not started before 2018, it can be assumed that the wrong drifts are recognized and a re-optimization with adjusted drifts delivers better results.

Other parameters influencing the investment decisions are the interest rate, the depreciation method, the method used to determine external investments and the existing portfolio of the considered operator. Lower interest rates and a straight-line depreciation increase the incentives of investments compared to higher interest rates or accelerated depreciation methods. The choice of these two parameters can be used to consider the willingness to take risks of the considered investor. If investments by competitors are disregarded, the amount of investments of the considered operator decreases. The considered operator behaves like a monopolist: he builds fewer new units as the amount required to satisfy the demand. As a consequence, a shortage of supply occurs during several hours a year, and the electricity price is set by the VoLL. An existing portfolio also decreases investment incentives, as the construction of new units reduces the contribution margins of existing units.

Despite of the deregulation of the electricity market, political decisions still have an important impact on the electricity market. Currently, there are two political issues, which significantly influence the profitability of new investments: the nuclear-phase out and the emission allowance allocation scheme. Concerning the nuclear phase-out, we have examined three scenarios: a nuclear-phase out as it was planned by the red-green government in 2002, an extension of the operational life of nuclear units by ten years, and an unrestricted extension of the operational life. Our tests have shown that the consequence of a delay or an annulment of the nuclear phase-out are postponed or annulled investments. A delayed or scrapped phase-out decreases the investment incentives for new entrants to the market, and hence it impedes a higher level of competition.

Following our tests, the emission allowance allocation scheme does not only influence the timing and the amount of new units, but also the preference for the type of new units. Currently, three different options concerning the allocation of emission allowances are discussed: a free allocation based on a fuel-specific benchmark, a free allocation based on a fuel-independent benchmark, and a full auctioning of emission allowances. A free allocation of emission allowances results in large windfall profits for the generation companies. Hence, investments in new units are more profitable and significantly more units are built than in the case where emission allowances must be bought. In our tests, the currently applicable fuel-dependent benchmark is the only scheme in which coal units are preferred to CCGT units.

In the last section of this chapter, we examined the profitability of coal-fired CCS units. Due to the difficulties of simulating carbon prices, e.g., the lack of a long history and a changing legal framework, we used the LP-based electricity market model to determine future carbon prices based on fundamental data. As the determination of carbon prices with the LP-model is based on several simplifications, e.g., the non-consideration of banking and the non-consideration of other countries, we have used a weighted average of model prices and expected carbon prices. While the resulting carbon prices were in most years below the expected prices, they were high enough to make the CCS technology profitable. On average, coal-fired CCS units were the preferred choice. Depending on the fuel price development, also new CCGT units were built, while non-CCS coal units were never built. Due to the consideration of the German market only and the non-consideration of banking, the construction of a single new unit can have a disproportionately high influence on the carbon price in our model. Nevertheless, we think that our results are interesting, as they indicate that CCS units will become profitable.

Chapter 6

Conclusions

Electricity is the basis for almost all economic activities nowadays. A reliable and cost-efficient supply of electricity is a decisive competitive factor for an economy. To ensure an adequate supply, the investment decisions of electricity generation companies are of great importance.

With the deregulation of the electricity market, the framework conditions for these investment decisions changed significantly. While generation companies have always been exposed to cost risks related to uncertainty about future fuel prices, the competition introduced by the deregulation has added two other forms of risk: price risk and volume risk. These risks depend on multiple factors like fuel and carbon price development, development of electricity demand and supply, and political decisions. Due to the competition, generation companies are no longer able to pass the whole risks to the consumers. Hence, the need for planning and decision support tools that are able to consider these risks arises.

In Germany, a significant amount of new power plants is required in the next years to replace old units, and—if the nuclear phase-out is not scrapped—to replace nuclear units. For most investment projects, two different options are discussed: coal units and CCGT units. While the local population and environmental organizations prefer the more ecologically CCGT units, generation companies often opt for the—in their opinion—more economically advantageous coal units. The main argument against CCGT units is the high risk related to future gas prices.

The aim of this thesis was to provide a model for power plant valuation with a focus on an adequate representation of fuel and carbon price uncertainty. Thereby, the interrelationship between fuel and carbon prices, electricity demand and supply on the one side and electricity prices and electricity production on the other side should be considered.

Summary

Basis for any method that is used to evaluate the profitability of power plants are the expected contribution margins for the new units. These contribution margins depend mainly on fuel and carbon prices, electricity prices and electricity production. As these parameters are highly interdependent, we argue that a fundamental electricity market model is the appropriate choice to determine the contribution margins.

After a general introduction to electricity markets in Chapter 1, we have presented several fundamental electricity market models with different levels of detail in Chapter 2. The least detailed model is the model based on the LDC. Due to its fast running times, it is best suited for a power plant valuation model that requires the calculation of contribution margins for a huge amount of different scenarios. A disadvantage of an LDC-based model is that it cannot consider technical restrictions of power plants. A consequence of this limitation is a lower standard deviation of the model's prices compared to the standard deviation of real electricity prices. To get a better approximation of the real prices, we used two different functions to adjust model prices: a mark-up function applied in times of scarcity, and a function that preserves average prices, but increases high prices while it decreases low prices. This function aims to simulate the effect of start-up costs and avoided start-up costs, respectively.

Compared to LDC models, fundamental electricity market models based on an LP formulation allow a much more detailed description of the electricity market. Due to this detailed description, running times get prohibitively high for a usage in an investment model. To reduce the complexity of such models, different simplifications, e.g., aggregation or the restriction to some typical days, are commonly used. But even with these simplifications, the computational burden is significantly higher than the one of an LDC model. Nevertheless, we examined such LP-based models, as they can also be used to determine carbon prices based on fundamental data. A comparison between the prices of both models and observed EEX spot-prices showed a good approximation of average electricity prices. As the investment alternatives we consider, coal units and CCGT units, are at least mid load units, we argue that a good approximation of average prices is more important than an exact approximation of the standard deviation of the prices. Hence, the LDC and LP models that we tested are suited to determine the contribution margins of new units for an investment model.

In a deregulated market, a power plant investment problem consists of different decisions: besides the amount and the type of the new units, the timing of the investments must be chosen. The option to postpone an investment can have a significant value, especially if new information, e.g., about fuel or carbon prices, becomes available in later years. The state of the art approach to tackle such problems is the real options approach. A common method to solve a real options problem is SDP.

In Chapter 3, we described how SDP can be used to optimize power plant investments taking into account fuel and carbon price uncertainty. We thereby took the position of a single, decentralized decision maker. Compared to similar models assuming a centralized decision maker, this choice required us to add a module that models investments of competitors. We opt for this more complicated perspective, as we do not believe that under current market conditions, both approaches necessarily lead to the same solution.

The main issue with an SDP formulation of the power plant investment problem is the "curse of dimensionality", which limits the usage of SDP to low dimensional problems. To be able to solve our power plant investment problem, several simplifications need to be made. The application of temporal aggregation, a common technique in SDP, allows us to significantly reduce the state space. However, it allows investment decisions only every four years,[1] and it introduces another problem: fuel prices now evolve stochastically within the period covered by a state. As the contribution margins based on average fuel prices do not correspond to the average contribution margins considering all fuel price scenarios associated to a state, we select several representative fuel price scenarios for every year of a state. The selection of these scenarios is done with the fast forward selection, a method commonly applied in stochastic programming.

However, the simplified problem is still computationally intractable, if a wide variety of fuel and carbon price scenarios should be considered. Motivated by the fact that for the unit commitment decisions between coal and CCGT units the cost difference between these units is decisive, we introduced another form of aggregation. Instead of explicitly saving the coal price, the gas price and the carbon price in our SDP model, we save the cost difference in form of the climate spread.

---

[1]The four years correspond to the construction time of new units that we used in our tests. To be exact, the proposed formulation allows the construction of new units every t' years, where t' corresponds to the maximum construction time of one of the considered investment alternatives.

The state space of this aggregated problem is much smaller compared to the state space of the original problem. Nevertheless, the size of the state space might still be problematic, especially if an LP-based electricity market model is used to determine the contribution margins for each state. Based on the idea that we do not need an exact value function for all states, we use an adaptive grid method to solve the problem. Thereby, we start with a coarse grid, consisting of a small subset of the state space. We calculate the contribution margins only for the grid points. The value of non-saved states, which may be reached during the optimization, is approximated by an interpolation based on the value of the grid points. For this reason, we refer to this model as interpolation-based stochastic dynamic programming (ISDP) model.

During the optimization, the grid is iteratively refined by splitting some of its hyperrectangles. The selection of the hyperrectangles that are split is based on the priority of each hyperrectangle determined by a splitting criterion. We presented three kinds of splitting criteria: value function based criteria aiming to refine the grid where the approximation error is the greatest, a policy based criterion, refining the grid around the optimal solution of the approximated problem, and a combination of the before mentioned approaches. It turned out that the combination of both approaches works best for our problem.

The application of the adaptive grid method allows us to consider a wide variety of fuel and carbon price scenarios. Based on a Monte Carlo simulation of these prices, we define an interval of climate spreads that should be covered by the model, and choose the boundary points of the grid correspondingly. Compared to similar models, the amount of considered fuel and carbon price scenarios of our model is much greater.

We compared the quality of the solutions of our ISDP model with the quality of the solutions of two alternatives: a deterministic approach based on expected fuel prices, and a normal SDP model. The quality of the solutions of both stochastic models is significantly better than the quality of the solution obtained by the deterministic model. The higher the uncertainty, the greater is the advantage of the stochastic models. In order to compare the ISDP model with a normal SDP model, we had to restrict the size of the considered test problems such that the SDP model is still able to solve them. For these test problems, the performance of both models was similar concerning solution quality and running times. However, the ISDP approach has two advantages compared to the normal SDP approach. First, it can be used to solve larger problems. Second, the ISDP approach is also faster for smaller problems, if an LP-based fundamental electricity market model is used instead of the LDC model.

Despite the advantages of the ISDP approach compared to to a normal SDP approach, the ISDP model still suffers from two limitations of the SDP approach: first, it allows new decisions only every four years, and secondly, even though it scales better than a normal SDP approach, it is still restricted to low dimensional problems. For this reason, we examined an approach that promises to break the curse of dimensionality: Approximate dynamic programming (ADP). Instead of explicitly saving the value of each state, ADP approximates the value of a state based on value function approximations depending only on some of a state's attributes. This approach has two advantages: first, the memory required to save the value function approximations is significantly lower than the memory required to save the value of all states in SDP. Secondly, it allows to approximate the value of states that have never been visited during the optimization. Hence, a complete enumeration of the state space can be avoided.

In Chapter 4, we examined how the power plant investment problem can be formulated and solved using ADP. We proposed to approximate the value of a state based on piecewise-linear value function approximations that depend on the number of new units already built and the number of new units under construction. To account for different fuel price scenarios and the amount of electricity demand that is not covered by competing units, we introduced the climate spread and the residual demand clusters. For each of these clusters, we use separate value function approximations.

A comparison of the solutions of the ADP model with those of the ISDP model showed for our test cases a slightly better solution quality of the ADP model. This is not surprising, as the ADP model allows annual investment decisions. In test cases with deterministic fuel prices, the ADP model was able to find the optimal solution.

Despite these good results, the ADP approach has several drawbacks. First, it requires a lot of parameter tunning. The wrong choice of a single parameter can result in an algorithm that does not work. Secondly, the approach cannot be used to consider an additional investment alternative. The reason is the requirement of independence of the attributes used for the value function approximations. For the considered problem, we could circumvent this problem with the introduction of the residual demand clusters. However, this method cannot be extended to more than two investment alternatives.

In Chapter 5, we applied our ISDP model to several case studies. Based on the assumption of an extension of the useful life of nuclear units by ten years, and a full auctioning of emission allowances, the case study showed that incentives for investments in new units are low. No new units are built before 2022. In most cases, CCGT units are preferred to coal units.

We used the base scenario to analyze the impact of different parameters on the solution of our model. We first examined the influence of different parameter choices concerning the stochastic process used for the fuel and carbon price simulation. It turned out that the higher the uncertainty and the higher the fuel and carbon price level, the more profitable are the investments in new units. This counterintuitive result is caused by the way electricity prices are set. If the price setting unit is less efficient than the new units, its costs rise more than those of the new units of the same technology. Hence, also electricity prices rise more than the costs of these new units.

Furthermore, we showed that the optimal decision depends on the portfolio of an investor. A generation company with an existing portfolio has less incentives to invest than a new entrant to the market—except when the company with the existing portfolio knows that if it does not invest, a competitor will invest. Moreover, our tests demonstrated that the non-consideration of competitors results in a lower degree of investments and in delayed investments.

We also studied the impact of political decisions on investments. A delay or an annulment of the nuclear phase-out significantly reduces the incentives to invest in new units. Besides the nuclear phase-out, we compared different emission allowance allocation schemes: a free allocation based on the current fuel-specific benchmark, a free allocation based on a fuel-independent benchmark and a full auctioning of the emissions. The only scheme under which coal units are preferred to CCGT units is the fuel-specific benchmark. The windfall profits obtained through the free allocation of emission allowances are an important incentive to invest in new units.

Finally, we evaluated the profitability of coal units with the carbon capture and storage (CCS) technology. The profitability of these units depends mainly on future carbon prices. To be able to determine carbon prices based on fundamentals, we used our LP-based electricity market model for this case study. Even though carbon prices tended to be slightly lower than expected, coal-fired CCS units were in most cases the preferred choice in this case study.

Main Contributions of this Thesis

The main contribution of this thesis is the development of methods allowing to combine fundamental electricity market models and SDP while considering a wide range of different fuel and carbon price scenarios. Compared to other approaches combining fundamental electricity market models and SDP, our methods allow for a more detailed representation of fuel and carbon price uncertainty. To the best of the author's knowledge, the ISDP approach using an adaptive grid method as well as the ADP approach have been applied for the first time to the power generation expansion problem.

The most important advantage of the ISDP approach is the reduction of the state space. Compared to a normal SDP approach, significantly less states need to be considered. This yields two benefits: First, memory requirements are lower. Hence, greater problems, modeling other aspects in more detail, can be solved. For example, the consideration of external investments—a further difference of our model to similar models in literature—was only possible in our ISDP model, but not in the standard SDP model that we used for the comparison of both approaches. Secondly, due to the reduced state space, less calculations of contribution margins for new units are required. This is especially important when an LP or an MILP model is used to determine the contribution margins. For our case study evaluating the profitability of CCS units, the running time was, despite a parallel implementation using eight cores, about a week. With our ISDP model, the contribution margins of about 14,000 states have been calculated. A normal SDP model based on the same fuel price intervals, would have consisted of about 475,000 states, and hence, it would have required a significantly longer running time.

A further method we have introduced to reduce the complexity of the problem is the aggregation of coal, gas and carbon prices to the climate spread. While this aggregation also significantly reduces the state space, the solutions based on this representation are only slightly worse than those based on an explicit representation of all three prices. However, it must be stressed that this technique is problem specific, as we defined the climate spread as the cost difference between coal and CCGT units. Hence, this approach will probably not work if other investment alternatives, e.g., lignite units, are additionally considered.

The more detailed representation of the electricity market is used, the greater is the advantage of the methods we proposed compared to normal SDP approaches. While this thesis focused on the comparison between coal and CCGT units, at least the adaptive grid method used by the ISDP model can also be applied to consider other investment alternatives, e.g., large offshore wind parks.

Concerning the ADP formulation of the power plant investment model, the first contribution of this thesis is the modeling of the problem. Compared to a normal SDP approach, the choice of the problem representation is more difficult in ADP. We have used an example to illustrate the effects of an inappropriate representation: in this case, the model was not able to capture the value of postponing an investment. We have introduced several problem-specific techniques to improve the algorithm that we use to solve the ADP problem. It turned out that these techniques, the initial update of neighboring value function approximations and the update of value function approximations with the same residual demand cluster, are very important for the performance of the algorithm.

Concerning the LP-based fundamental electricity market model used to determine the contribution margins of new units, we addressed several problems caused by the simplification of the model, which are disregarded in other models. We proposed an alternative formulation of the minimum up-time and minimum down-time constraints, which also holds when units are aggregated to unit groups. Moreover, this formulation is more efficient than the one commonly used. In addition, we discussed several problems related to fundamental electricity market models based on typical days, and proposed alternative formulations that take these problems into account. Especially for an (approximately) realistic storage operation, it is important to consider the proposed adjustments.

Applicability and Limitations

The models described in this thesis were developed from the point of view of an electricity generation company. The aim was to provide a model that can be used as a decision-support tool for long-term power plant investment planning under fuel and carbon price uncertainty. As the profits of the units of the considered operator depend, amongst others, on the decisions of competitors, we included a module determining the investments of competitors. Depending on the parameters used for this module, more or less aggressive competitors can be simulated.

Even though our model is formulated from the point of view of a single company, it might also be interesting for regulatory authorities and politicians to test the impact of different legal conditions. For example, the effect of different emission allowance allocation schemes can be tested, as we have done it in one of the case studies.

As the power generation expansion problem is a very complex problem, there are a wide variety of different approaches, focusing on different aspects of the problem. As such, each of these approaches has its own strengths and weaknesses. For example, even though we included an investment module modeling investments of competitors, game-theoretic approaches are better suited to model the strategic behavior of different generation companies.

The focus of our model was on a strategic level. Our model disregards several soft factors, which also influence investment decisions. For example, we do not consider a possible reservation against gas fired units due to the fear to be dependent on Russia. We neither consider the resistance of environmental organizations and parts of the population against coal units, even though this resistance stopped several planned coal units recently (e.g., the planned coal units in Ensdorf and Emden (FAZ [FAZ09])).

In summary, we believe that due to the ability to consider a wide variety of different fuel and carbon price scenarios, our model is well suited for long term investment planning. However, to incorporate other aspects of the problem, it should be used in conjunction with models focusing on these aspects.

Directions for Future Research

The focus of this thesis was on an adequate representation of fuel and carbon price uncertainty. Other kinds of uncertainty, e.g., uncertainty about the legal framework or uncertainty about future electricity demand, have been disregarded or have only been addressed with a comparison of different scenarios. Our ISDP model can be easily extended to consider at least some of these uncertainties explicitly. The consideration of an additional kind of uncertainty would just require the introduction of an additional state attribute. For example, one may model electricity demand depending on the past electricity prices: high electricity prices may decrease electricity demand, while lower prices stimulate the demand.

As we took the perspective of a single operator in the market, we had to consider the decisions of competitors. For this purpose, we described two relatively simple approaches how these decisions can be modeled. Future research may concentrate on an integration of game-theoretic approaches into the investment model.

The combination of an LP-based electricity market model with the ISDP investment model allowed us to determine carbon price based on fundamentals. As we have discussed, this method is based on several simplifications. First, we only considered a part of the market. This restriction can be circumvented by modeling the whole European electricity market. Secondly, our approach does not consider banking of emission allowances. A possibility would be to add another attribute to a state in the investment model. This attribute could be used to count the surplus or shortage of emission allowances, which can be added or deduced from the amount of allowed emissions in the subsequent years.

Another possible direction for future research lies in the objective function used for the investment problem. The models presented in this thesis maximize the expected value of the investment decisions. One may extend this approach by considering also the risk related to the decisions. A possibility would be to integrate risk measures like the value at risk or the conditional value at risk. It would also be possible to extend the problem formulation to a multi-objective optimization problem, using the maximization of the expected profits as one objective, and the minimization of the associated risks as another objective.

Nomenclature

Abbreviations

ACVD	Average Corner Value Difference
ADP	Approximate Dynamic Programming
AE	Approximation Error
CAES	Compressed Air Energy Storage
CCGT	Combined Cycle Gas Turbine
CCS	Carbon Capture and Storage
CDS	Clean Dark Spread
CSS	Clean Spark Spread
DCF	Discounted Cash Flow
DP	Dynamic Programming
ECX	European Climate Exchange
EEX	European Energy Exchange
EU ETS	European Union Emission Trading System
EUA	European Union Allowance
GBM	Geometric Brownian Motion
ISDP	Interpolation-based Stochastic Dynamic Programming
LCEP	Levelized Costs of Electricity Production
LDC	Load–Duration Curve

LP	Linear Programming
MAE	Mean Average Error
MILP	Mixed-Integer Linear Programming
NAP	National Allocation Plan
NPV	Net Present Value
PIOS	Probability Interpolated Optimal Solution
RMSE	Root Mean Squared Error
RSI	Residual Supply Index
SDC	Supply–Demand Curve
SDP	Stochastic Dynamic Programming
TSO	Transmission System Operator
VNL	Value Non-Linearity
VoLL	Value of Lost Load
WACVD	Weighted Average Corner Value Difference
WAE	Weighted Approximation Error
WVNL	Weighted Value Non-Linearity

Notation used for the Fundamental Electricity Market Model

Sets and Indices

$n \in \mathcal{N}$	Nodes
$t \in \mathcal{T}$	Timesteps
$u \in \mathcal{U}$	Units
$u \in \mathcal{U}^{\text{FastStart}}$	Fast starting units (Units that are able to start fast-enough to provide minute reserves if they are turned off.)
$u \in \mathcal{U}^{\text{Storage}}$	Electricity storages

Variables

F	Fuel consumption [MWh]
FC	Fuel consumption costs [€]
OC	Operating and maintenance costs [€]
P	Electricity production [MWh]
p^{Online}	Capacity online [MW]
$p^{\text{Res}-}$	Negative reserves [MW]
$p^{\text{ResNonSpin}+}$	Positive non-spinning reserves [MW]
$p^{\text{ResSpin}+}$	Positive spinning reserves [MW]
p^{ShutDown}	Capacity shut-down [MW]
p^{StartUp}	Capacity started [MW]
SC	Start-up costs [€]
TC	Total costs [€]
V	Storage content [MWh]
W	Electricity used for storage charging [MWh]

Parameters

α	Fuel consumption parameter [MWh$_{\text{fuel}}$/MW$_{\text{elec}}$] for being online
β	Fuel consumption parameter [MWh$_{\text{fuel}}$/MWh$_{\text{elec}}$] for producing electricity
η	Storage efficiency [%]
γ	Operating and maintenance costs [€/MWh]
κ	CO$_2$ emissions [t/MWh]
π_n	Occurrence probability of n
$\psi_{n' \to n}$	Transition probability from n' to n
ρ	Availability factor
θ	Additional fuel usage for start-ups [MW$_{\text{fuel}}$/MW$_{\text{elec}}$]
c^{abrasion}	Abrasion costs [€/MW]
c^{CO_2}	Carbon costs [€/t]
$c^{\text{Fuel}}_{f(u)}$	Fuel costs depending on fuel f used by unit u [€/MWh]
$d^{\text{Res}-}$	Negative reserve demand [MW]
$d^{\text{ResNonSpin}+}$	Positive non-spinning reserve demand [MW]
$d^{\text{ResSpin}+}$	Positive spinning reserve demand [MW]
d	Duration of a timestep [h]
em^{max}	Maximum allowed CO$_2$ emissions [t]
f_t	Frequency of timestep t
l	Load [MW]
m_t	Number of times the representative day to which t belongs is repeated
p^{max}	Maximum electricity production [MW]

$p_u^{\min}$	Minimum (relative) electricity production.
t^{MinDown}	Minimum downtime
t^{MinUp}	Minimum operation time
t^{PerDay}	Timesteps per day
$v^{\max}$	Maximum storage content [MWh]
$w^{\max}$	Maximum storage charging [MW]

Notation used for the ISDP Investment Model

Indices

$f \in \mathcal{F}$	Fuels for which stochastic prices are used in the model, e.g., *coal* or *gas*
$s \in \{0, 1, \ldots, N-1\}$	Stage
$t \in \{0, 1, \ldots, T-1\}$	Year
$u \in \mathcal{U}$	Types of new units that can be built, e.g., *CCGT unit* or *coal unit*

Variables and parameters

β	Discount factor
$a \in \Gamma(x)$	Action (depending on state x)
$c^{\text{Invest}}(a)$	Investment costs depending on action a
$F(x, a)$	Payoff function depending on state x and action a
$P(x_{s+1} \mid x_s, a_s)$	Conditional probability of state x_{s+1} under current state x_s and action a_s
$R(x)$	Reward for being in state x
$T(x, a)$	Transition function from state x taking action a
t_s^1	First year of stage s
T^s	Number of years per stage
$V(x)$	Value function for state x
$\tilde{V}(x)$	Approximated value function for (the non-saved) state x
$x_s \in \mathcal{X}_s$	State at stage s
x^f	Vector of attributes of a state characterizing the fuel price scenarios

x^u	Vector of attributes of a state containing the number of new units built by the considered operator $(x^{u^{\mathrm{Own}}})$ and those built by competitors $(x^{u^{\mathrm{Ext}}})$ for each type of unit
$x^{u^{\mathrm{Ext}}}$	Vector of attributes of a state containing the number of new units built by competitors
$x^{u^{\mathrm{Own}}}$	Vector containing the attributes of a state counting the number of new units built by the considered operator
$\mathcal{X}_s^{\mathrm{NonSaved}}$	Set of non-saved states at stage s
$\mathcal{X}_s^{\mathrm{Saved}}$	Set of saved states at stage s

Variables and parameters related to the determination of the contribution margins

$C_t^{\mathrm{Fuel}}(x)$	Set of all fuel price scenarios associated to state x in year t
$C_t^{\mathrm{Fuel,NonRep}}(x)$	Set of non-representative fuel prices in year t for state x
$C_t^{\mathrm{Fuel,Rep}}(x)$	Set of representative fuel prices in year t for state x
$P(c^{\mathrm{Fuel}} \mid x, t)$	Probability of fuel price scenario c^{Fuel} for state x and year t
$R^{\mathrm{Prod}}(x^u, c^{\mathrm{Fuel}}, t)$	Contribution margin depending on units x^u, fuel prices c^{Fuel} and year t
$\widetilde{R}^{\mathrm{Prod}}(x^u, c^{\mathrm{Fuel}}, t)$	Approximated contribution margin depending on units x^u, fuel prices c^{Fuel} and year t

Notation used for the ADP Investment Model

α_n	Stepsize used for the value function update in iteration n
β	Discount factor
π	Policy specifying an action for each state
$\phi^{u_{t'}}(x_t)$	Residual demand cluster and climate spread cluster for units of type u in year t' depending on state x_t
$a_t \in \Gamma(x_t)$	Decision in year t depending on state x_t
$c(a_t)$	(Investment) costs of decision a_t
c_t^f	Price of fuel (or carbon) f in year t
csc_t	Climate spread cluster in year t
$F(x_t, a_t)$	Payoff function depending on state x_t and decision a_t
$R_t(x_t)$	Reward for being in state x_t in year t
rdc_t^u	Residual demand cluster for units of type u in year t
t_u^{Constr}	Construction time of units of type u
$T^w(x_t^a, w_{t+1})$	Transition function from post-decision state x_t^a to the succeeding pre-decision state depending on realization w_{t+1} of the stochastic process
$T^a(x_t, a_t)$	Transition function from pre-decision state x_t to the succeeding post-decision state when decision a_t is applied
$\hat{v}_{t,u_{t'}}$	Marginal value of an additional unit of type u in year t', where t is the current year
$\tilde{V}_t(x_t^a)$	Value function approximation of post-decision state x_t^a
$\tilde{V}_{t,u_{t'}}\left(x_{t'}^u \mid \phi^{u_{t'}}(x_t)\right)$	Piecewise-linear value function approximation of the number of new units of type u in year t' depending on the climate spread and residual demand clusters of the current state x_t. Thereby, t denotes the current year. If $t' = t + t^{\text{Constr}}$, then this value corresponds to expected value until

the end of the optimization horizon, otherwise only to the value in t'.

$\tilde{V}_{t-1}^m \leftarrow U^V(\tilde{V}_{t-1}^{m-1}, x_{t-1}^m, \hat{v}_t^m)$ Update of value function $\tilde{V}_{t-1}^m$ at iteration m and year $t-1$ depending on the value function approximation during the previous iteration $\tilde{V}_{t-1}^{m-1}$, the state $x_{t-1}^{a,m}$, and the observed value $\hat{v}_t^m$

w_t Realization of the stochastic process in year t

x_t Pre-decision state in year t

x_t^a Post-decision state in year t

x_t^u Number of new units of type u in year t

Bibliography

[Aro+08]   Alexandre Vasconcelos Aronne, Haroldo Guimarães Brasil, and Ivan Dionysio Aronne. *Valuation of Investments in Flexible Power Plants: A Case Study in the Brazilian Power Market*. Paper presented at the 12th Annual International Conference on Real Options: Theory Meets Practice, Rio de Janeiro, Brasil. July 2008. URL: http://www.realoptions.org/papers2008/Aronne%20Alexandre%20-%20AronneRO.pdf (visited on 12/16/2009).

[ATK09]   A.T. Kearney. *Von der Finanzkrise zur Energiekrise? - Die Auswirkungen der Finanzkrise auf die Energiewirtschaft*. Studie. Berlin, Feb. 2009.

[Bag02]   Joachim Bagemihl. "Optimierung eines Portfolios mit hydrothermischem Kraftwerkspark im börslichen Strom- und Gasterminmarkt". Dissertation. Institute of Energy Economics and the Rational Use of Energy, Universität Stuttgart, 2002.

[Bar02]   Martin T. Barlow. "A Diffusion Model for Electricity Prices". In: *Mathematical Finance* 12.4 (2002), pp. 287–298. DOI: 10.1111/j.1467-9965.2002.tb00125.x.

[Bar+08]   Rüdiger Barth, Ansgar Geiger, Bernhard Hasche, and Jan Mohring. *NetMod – Reduzierte Modelle komplexer elektrischer Netze mit verteilten Energieerzeugungssystemen – Validierung zur Einsatzoptimierung im überregionalen Netz*. Projektbericht 4.3. 2008.

[Bel57]   Richard Bellman. *Dynamic Programming*. Princeton, N.J.: Princeton University Press, 1957.

[Ben+02]   Michele Benini, Mirko Marracci, Paolo Pelacchi, and Andrea Venturini. "Day-ahead market price volatility analysis in deregulated electricity markets". In: *Power Engineering Society Summer Meeting, 2002 IEEE*. Vol. 3. 2002, pp. 1354–1359. DOI: 10.1109/PESS.2002.1043596.

[Ber05]   Dimitri P. Bertsekas. *Dynamic Programming and Optimal Control*.
            3rd ed. Belmont, Mass.: Athena Scientific, 2005.

[Bes+95]   Hendrik Bessembinder, Jay F. Coughenour, Paul J. Seguin, and Mar-
            garet Monroe Smoller. "Mean Reversion in Equilibrium Asset Prices:
            Evidence from the Futures Term Structure". In: *The Journal of Finance*
            50.1 (1995), pp. 361–375.

[BL02]   Hendrik Bessembinder and Michael L. Lemmon. "Equilibrium Pric-
            ing and Optimal Hedging in Electricity Forward Markets". In: *The
            Journal of Finance* 57.3 (2002), pp. 1347–1382. DOI: 10.1111/1540-
            6261.00463.

[BL92]   Derek W. Bunn and Erik R. Larsen. "Sensitivity of reserve margin to
            factors influencing investment behaviour in the electricity market of
            England and Wales". In: *Energy Policy* 20.5 (May 1992), pp. 420–429.
            DOI: 10.1016/0301-4215(92)90063-8.

[BL97]   John R. Birge and François Louveaux. *Introduction to Stochastic Pro-
            gramming*. Springer Verlag, 1997.

[BMU07]   Bundesministerium für Umwelt, Naturschutz und Reaktorsicherheit
            (BMU). *Revidierter Nationaler Allokationsplan 2008–2012 für die Bun-
            desrepublik Deutschland*. Berlin, Feb. 2007.

[BMU09]   Bundesministerium für Umwelt, Naturschutz und Reaktorsicherheit
            (BMU). *15. Klimakonferenz (Kopenhagen) – Verhandlungsposition der Eu-
            ropäischen Union*. Website. Dec. 2009. URL: http://www.bmu.de/15_
            klimakonferenz/doc/45239.php (visited on 12/02/2009).

[BMW09]   Bundesministerium für Wirtschaft und Technologie (BMWi). *En-
            twicklung von Energiepreisen und Preisindizes*. 2009. URL: http://www.
            bmwi.de/BMWi/Navigation/Energie/energiestatistiken,did=
            180914.html (visited on 05/27/2009).

[Bol86]   Tim Bollerslev. "Generalized Autoregressive Conditional Het-
            eroskedasticity". In: *Journal of Econometrics* 31.3 (1986), pp. 307–327.
            DOI: 10.1016/0304-4076(86)90063-1.

[Bor00]   Severin Borenstein. "Understanding Competitive Pricing and Mar-
            ket Power in Wholesale Electricity Markets". In: *The Electricity Journal*
            13.6 (July 2000), pp. 49–57. DOI: 10.1016/S1040-6190(00)00124-X.

[Bot03]   Audun Botterud. "Long-Term Planning in Restructured Power Systems: Dynamic Modelling of Investments on New Power Generation under Uncertainty". PhD thesis. Faculty of Information Technology, Mathematics and Electrical Engineering, Norwegian University of Science and Technology (NTNU), Dec. 2003. URL: http://www.diva-portal.org/ntnu/abstract.xsql?dbid=48 (visited on 12/16/2009).

[Bot+05]   Audun Botterud, Marija D. Ilic, and Ivar Wangensteen. "Optimal investments in power generation under centralized and decentralized decision making". In: *Power Systems, IEEE Transactions on* 20.1 (2005), pp. 254–263. DOI: 10.1109/TPWRS.2004.841217.

[Box+08]   George E. P. Box, Gwilym M. Jenkins, and Gregory C. Reinsel. *Time Series Analysis: Forecasting and Control*. 4th ed. Hoboken, N.J.: John Wiley, June 2008.

[BP09]   BP. *BP Statistical Review of World Energy June 2009*. 2009. URL: http://www.bp.com/statisticalreview (visited on 12/17/2009).

[Bre+07]   Richard Brealey, Stewart C. Myers, and Franklin Allen. *Principles of Corporate Finance*. 9th ed. McGraw-Hill, 2007.

[BS73]   Fischer Black and Myron Scholes. "The Pricing of Options and Corporate Liabilities". In: *The Journal of Political Economy* 81.3 (1973), 637Ű–654.

[BS85]   Michael J. Brennan and Eduardo S. Schwartz. "Evaluating Natural Resource Investments". In: *The Journal of Business* 58.2 (Apr. 1985), pp. 135–157.

[BT09]   Eva Benz and Stefan Trück. "Modeling the price dynamics of CO_2 emission allowances". In: *Energy Economics* 31.1 (2009), pp. 4–15. DOI: 10.1016/j.eneco.2008.07.003.

[BT95]   Dimitri P. Bertsekas and John N. Tsitsiklis. "Neuro-dynamic programming: an overview". In: *Decision and Control, 1995., Proceedings of the 34th IEEE Conference on*. Vol. 1. 1995, pp. 560–564. DOI: 10.1109/CDC.1995.478953.

[BT96]   Dimitri P. Bertsekas and John N. Tsitsiklis. *Neuro-Dynamic Programming*. Belmont, Mass.: Athena Scientific, 1996.

[BUN06]   Bund für Umwelt- und Naturschutz (BUND). *Stellungnahme des BUND zum Entwurf des Nationalen Allokationsplans 2008 bis 2012.* May 2006. URL: http://www.bmu.de/files/pdfs/allgemein/application/pdf/nap_stellungnahme_bund.pdf (visited on 12/16/2009).

[Bur+04]   Markus Burger, Bernhard Klar, Alfred Müller, and Gero Schindlmayr. "A spot market model for pricing derivatives in electricity markets". In: *Quantitative Finance* 4 (Feb. 2004), pp. 109–122. DOI: 10.1088/1469-7688/4/1/010.

[Bur+07]   Markus Burger, Bernhard Graeber, and Gero Schindlmayr. *Managing Energy Risk: An Integrated View on Power and Other Energy Markets.* 1st ed. John Wiley, Nov. 2007.

[CA06]   Miguel Carrión and José Manuel Arroyo. "A Computationally Efficient Mixed-Integer Linear Formulation for the Thermal Unit Commitment Problem". In: *IEEE Transactions on Power Systems* 21.3 (2006), pp. 1371–1378. DOI: 10.1109/TPWRS.2006.876672.

[CA09]   Karan Capoor and Philippe Ambrosi. *State and Trends of the Carbon Market 2009.* The World Bank. Washington, D.C., May 2009.

[Cai07]   Caisse des Dépôts. *Tendances Carbone – Methodology – Version 3.* The European carbon market monthly bulletin. Sept. 2007. URL: http://www.caissedesdepots.fr/fileadmin/PDF/finance_carbone/document_methodologie_tendances_carbone_en_v4.pdf (visited on 12/16/2009).

[CF09]   CDU/CSU and FDP. *Wachstum. Bildung. Zusammenhalt.* Koalitionsvertrag zwischen CDU, CSU und FDP 17. Legislaturperiode. Berlin, Oct. 2009.

[Chu+01]   Angela S. Chuang, Felix Wu, and Pravin Varaiya. "A game-theoretic model for generation expansion planning: problem formulation and numerical comparisons". In: *Power Systems, IEEE Transactions on* 16.4 (2001), pp. 885–891. DOI: 10.1109/59.962441.

[CON+08]   CONSENTEC, EWI, and IAEW. *Analyse und Bewertung der Versorgungssicherheit in der Elektrizitätsversorgung.* Untersuchung im Auftrag des Bundesministerium für Wirtschaft und Technologie (BMWi). May 2008.

[CS05]   Peter Cramton and Steven Stoft. "A Capacity Market that Makes Sense". In: *The Electricity Journal* 18.7 (Sept. 2005), pp. 43–54. DOI: 10.1016/j.tej.2005.07.003.

[Cze+03]   Clemens Czernohous, Wolf Fichtner, Daniel J. Veit, and Christof Weinhardt. "Management decision support using long-term market simulation". In: *Information Systems and E-Business Management* 1.4 (Nov. 2003), pp. 405–423. DOI: 10.1007/s10257-003-0021-3.

[Das+09]   George Daskalakis, Dimitris Psychoyios, and Raphael N. Markellos. "Modeling CO_2 emission allowance prices and derivatives: Evidence from the European trading scheme". In: *Journal of Banking & Finance* 33.7 (July 2009), pp. 1230–1241. DOI: 10.1016/j.jbankfin.2009.01.001.

[Dav97]   Scott Davies. "Multidimensional Triangulation and Interpolation for Reinforcement Learning". In: *Advances in Neural Information Processing Systems* 9 (1997), pp. 1005–1011.

[DE08]   Henry Dannenberg and Wilfried Ehrenfeld. *Prognose des CO_2-Zertifikatepreisrisikos*. IWH-Diskussionspapiere 5. Halle Institute for Economic Research, May 2008. URL: http://ideas.repec.org/p/iwh/dispap/5-08.html (visited on 12/16/2009).

[Den05]   Shi-Jie Deng. "Valuation of investment and opportunity-to-invest in power generation assets with spikes in electricity price". In: *Managerial Finance* 31 (June 2005), pp. 95–115. DOI: 10.1108/03074350510769712.

[DEN08]   Deutsche Energie-Agentur GmbH (dena). *Kurzanalyse der Kraftwerks- und Netzplanung in Deutschland bis 2020 (mit Ausblick auf 2030)*. Studie. Berlin, Apr. 2008.

[Den99]   Shijie Deng. *Stochastic Models of Energy Commodity Prices and Their Applications: Mean-reversion with Spikes and Jumps*. Working Paper PWP-073. University of California Energy Institute, 1999.

[Des08]   Statistisches Bundesamt (Destatis). *Stromerzeugungsanlagen der Betriebe im Bergbau und im Verarbeitenden Gewerbe 2007*. Fachserie 4 Reihe 6.4. Wiesbaden, Sept. 2008.

[Des09a]   Statistisches Bundesamt (Destatis). *Index der Einfuhrpreise*. 2009. URL: https://www-genesis.destatis.de/genesis/online/ (visited on 12/02/2009).

[Des09b]   Statistisches Bundesamt (Destatis). *Stromabsatz und Erlöse der Elektrizitätsversorgungsunternehmen*. 2009. URL: https://www-genesis.destatis.de/genesis/online/ (visited on 12/17/2009).

[Des09c]   Statistisches Bundesamt (Destatis). *Verbraucherpreisindex für Deutschland*. Lange Reihen ab 1948, Oktober 2009. Wiesbaden, Nov. 2009.

[DEW+05]   DEWI, E.ON Netz, EWI, RWE Transportnetz Strom, and VE Transmission. *Energiewirtschaftliche Planung für die Netzintegration von Windenergie in Deutschland an Land und Offshore bis zum Jahr 2020 (dena Netzstudie)*. Endbericht. Köln: Deutsche Energie-Agentur GmbH (dena), Feb. 2005.

[DM90]   Christian Darken and John Moody. "Note on Learning Rate Schedules for Stochastic Optimization". In: *Proceedings of the 1990 conference on Advances in neural information processing systems 3*. Denver, Colorado, United States: Morgan Kaufmann Publishers Inc., 1990, pp. 832–838.

[DP94]   Avinash K. Dixit and Robert S. Pindyck. *Investment under Uncertainty*. Princeton University Press, 1994.

[EEX08]   European Energy Exchange (EEX). *Growth through security*. Annual Report 2008. Leipzig, 2008.

[EnB08]   EnBW AG. *Energy is a responsibility*. Annual Report. 2008.

[EON08]   E.ON AG. *Local. International*. Annual Report. 2008.

[EP07]   EWI and Prognos AG. *Energieszenarien für den Energiegipfel 2007*. Endbericht. Basel/Köln, Nov. 2007.

[Esc+02]   Álvaro Escribano, Juan Ignacio Peña, and Pablo Villaplana. *Modeling Electricity Prices: International Evidence*. Economics Series 08, Working Paper 02-27. Departamento de Economía, Universidad Carlos III, June 2002. URL: http://ideas.repec.org/p/cte/werepe/we022708.html (visited on 12/16/2009).

[Eur07]   The European Commission. *Emissions trading: Commission adopts decision on Italy's national allocation plan for 2008-2012*. Press Release IP/07/667. Brussels, May 2007. URL: http://europa.eu/rapid/pressReleasesAction.do?reference=IP/07/667 (visited on 12/16/2009).

[Fat07]   Bassam Fattouh. *The Drivers of Oil Prices: The Usefulness and Limitations of Non-Structural model, the Demand–Supply Framework and Informal Approaches*. EIB Papers, Volume 12 1. European Investment Bank, Economic and Financial Studies, June 2007.

[FAZ09]   Frankfurter Allgemeine Zeitung (FAZ). *Kohle und CO_2 – Je heißer der Dampf, desto weniger Kohlendioxid*. Dec. 2009. URL: http://www.faz.net/s/RubC5406E1142284FB6BB79CE581A20766E/Doc~E70816B95D7E4484F9451CD16C7747D43~ATpl~Ecommon~Scontent.html (visited on 12/20/2009).

[FN03]   Stein-Erik Fleten and Erkka Näsäkkälä. *Gas fired power plants: Investment timing, operating flexibility and CO_2 capture*. MPRA Paper 15716. Mar. 2003. URL: http://mpra.ub.uni-muenchen.de/15716/ (visited on 12/16/2009).

[For01]   Andrew Ford. "Waiting for the boom: a simulation study of power plant construction in California". In: *Energy Policy* 29.11 (Sept. 2001), pp. 847–869. DOI: 10.1016/S0301-4215(01)00035-0.

[For58]   Jay W. Forrester. "Industrial Dynamics: A Major Breakthrough for Decision Makers." In: *Harvard Business Review* 36.4 (1958), pp. 37–66.

[Gei+08]   Ansgar Geiger, Chris Heyde, Krzysztof Rudion, Tomasz Smieja, and Zbigniew A. Styczynski. *NetMod – Reduzierte Modelle komplexer elektrischer Netze mit verteilten Energieerzeugungssystemen – Entwicklung von Optimierungsverfahren für Auslegungs- und Betriebsführungsprobleme*. Projektbericht 4.2. Sept. 2008.

[Gem07]   Hélyette Geman. "Mean Reversion Versus Random Walk in Oil and Natural Gas Prices". In: *Advances in Mathematical Finance*. 2007, pp. 219–228. DOI: 10.1007/978-0-8176-4545-8_12.

[Gen+07]   Massimo Genoese, Frank Sensfuß, Dominik Möst, and Otto Rentz. "Agent-Based Analysis of the Impact of CO_2 Emission Trading on

Spot Market Prices for Electricity in Germany". In: *Pacific Journal of Operations Research* 3.3 (2007), 401Ű–423.

[Gen+08]   Massimo Genoese, Dominik Möst, Philip Gardyan, and Otto Rentz. "Impact of emission allocation schemes on power plant investments". In: *New methods for energy market modelling*. Karlsruhe: Universitätsverlag Karlsruhe, 2008, pp. 29–47.

[GH01]   John R. Graham and Campbell R. Harvey. "The theory and practice of corporate finance: evidence from the field". In: *Journal of Financial Economics* 60.2-3 (May 2001), pp. 187–243. DOI: 10.1016/S0304-405 X(01)00044-7.

[GK+03]   Nicole Gröwe-Kuska, Holger Heitsch, and Werner Römisch. "Scenario reduction and scenario tree construction for power management problems". In: *Power Tech Conference Proceedings, 2003 IEEE Bologna*. Vol. 3. 2003, p. 7. DOI: 10.1109/PTC.2003.1304379.

[Göb01]   Matthias Göbelt. "Entwicklung eines Modells für die Investitions- und Produktionsprogrammplanung von Energieversorgungsunternehmen im liberalisierten Markt". Dissertation. Fakultät für Wirtschaftswissenschaften, Universität Karlsruhe (TH), 2001.

[GP06]   Abraham P. George and Warren B. Powell. "Adaptive stepsizes for recursive estimation with applications in approximate dynamic programming". In: *Machine Learning* 65.1 (Oct. 2006), pp. 167–198. DOI: 10.1007/s10994-006-8365-9.

[GR06]   Hélyette Geman and Andrea Roncoroni. "Understanding the Fine Structure of Electricity Prices". In: *The Journal of Business* 79.3 (2006), pp. 1225–1261. DOI: 10.1086/500675.

[Gre09]   Greenpeace. *Mehrheit der Deutschen will den Atomausstieg*. Presseerklärungen zum Thema Atomkraft. Sept. 2009. URL: http://www.greenpeace.de/themen/atomkraft/presseerklaerungen/artikel/mehrheit_der_deutschen_will_den_atomausstieg/ (visited on 12/16/2009).

[Gri07]   Vanessa Grimm. "Einbindung von Speichern für erneuerbare Energien in die Kraftwerkseinsatzplanung – Einfluss auf die Strompreise der Spitzenlast". Dissertation. Fakultät für Maschinenbau, Ruhr-Universität Bochum, 2007.

[GS04]   Lars Grüne and Willi Semmler. "Using dynamic programming with adaptive grid scheme for optimal control problems in economics". In: *Journal of Economic Dynamics and Control* 28.12 (Dec. 2004), pp. 2427–2456. DOI: 10.1016/j.jedc.2003.11.002.

[GS90]   Rajna Gibson and Eduardo S. Schwartz. "Stochastic Convenience Yield and the Pricing of Oil Contingent Claims". In: *The Journal of Finance* 45.3 (1990), 959–Ű976.

[Han+09]   Aoife Brophy Haney, Tooraj Jamasb, and Michael G. Pollitt. *Smart Metering and Electricity Demand: Technology, Economics and International Experience*. Cambridge Working Papers in Economics 0905. Faculty of Economics, University of Cambridge, Feb. 2009.

[Har+68]   Peter E. Hart, Nils J. Nilsson, and Bertram Raphael. "A Formal Basis for the Heuristic Determination of Minimum Cost Paths". In: *Systems Science and Cybernetics, IEEE Transactions on* 4.2 (1968), pp. 100–107. DOI: 10.1109/TSSC.1968.300136.

[Här92]   Wolfgang Härdle. *Applied nonparametric regression*. Cambridge University Press, 1992.

[Hen+03]   Bruce Henning, Michael Sloan, and Maria de Leon. *Natural Gas and Energy Price Volatility*. Prepared for the Oak Ridge National Laboratory. Arlington, Virginia: American Gas Foundation, Oct. 2003.

[Hh+07]   Christian von Hirschhausen, Hannes Weigt, and Georg Zachmann. *Preisbildung und Marktmacht auf den Elektrizitätsmärkten in Deutschland – Grundlegende Mechanismen und empirische Evidenz*. Gutachten im Auftrag des Verbandes der Industriellen Energie- und Kraftwirtschaft e.V. 2007.

[HL05]   Frederick S. Hillier and Gerald J. Lieberman. *Introduction to Operations Research*. 8th ed. Boston: McGraw-Hill, 2005.

[HS09]   Matthias V. Hundt and Ninghong Sun. "Modelling thermal power plants as real options applying stochastic mixed-integer programming". In: *Conference Proceedings of the 6th International Conference on the European Energy Market (EEM09)*. Leuven, Belgium, May 2009.

[IO03]   International Energy Agency (IEA) and Organisation for Economic Co-operation and Development (OECD). *Power generation investment in electricity markets*. Paris: IEA and OECD, 2003.

[JM08]   Stine Grenaa Jensen and Peter Meibom. "Investments in liberalised power markets: Gas turbine investment opportunities in the Nordic power system". In: *International Journal of Electrical Power & Energy Systems* 30.2 (Feb. 2008), pp. 113–124. DOI: 10.1016/j.ijepes.2007.06.029.

[Jos08]   Paul L. Joskow. "Capacity payments in imperfect electricity markets: Need and design". In: *Utilities Policy* 16.3 (Sept. 2008), pp. 159–170. DOI: 10.1016/j.jup.2007.10.003.

[Kae+96]   Leslie Pack Kaelbling, Michael L. Littman, and Andrew W. Moore. "Reinforcement learning: a survey". In: *Journal of Artificial Intelligence Research* 4 (1996), pp. 237–285. DOI: 10.1.1.31.3460.

[Kag+04]   Argyris G. Kagiannas, Dimitris Th. Askounis, and John Psarras. "Power generation planning: a survey from monopoly to competition". In: *International Journal of Electrical Power & Energy Systems* 26.6 (July 2004), pp. 413–421. DOI: 10.1016/j.ijepes.2003.11.003.

[Kal39]   Nicholas Kaldor. "Speculation and Economic Stability". In: *The Review of Economic Studies* 7.1 (Oct. 1939), pp. 1–27.

[Kan09]   Takashi Kanamura. *A Classification Study of Carbon Assets into Commodities*. Working Paper. 2009. URL: http://papers.ssrn.com/sol3/papers.cfm?abstract_id=1332267 (visited on 12/16/2009).

[KL03]   Jussi Keppo and Hao Lu. "Real options and a large producer: the case of electricity markets". In: *Energy Economics* 25.5 (Sept. 2003), pp. 459–472. DOI: 10.1016/S0140-9883(03)00048-3.

[Kle+08]   Alexandre Klein, Julian Bouchard, and Sabine Goutier. *Generation capacity expansion under long-term uncertainties in the US electric market*. Paper presented at the 12th USAEE/IAEE North American Conference – Unveiling the Future of Energy Frontiers, New Orleans, Louisiana USA. Dec. 2008. URL: http://www.usaee.org/usaee2008/submissions/OnlineProceedings/Paper_IAEE_122008_Klein_Bouchard_Goutier.pdf (visited on 12/16/2009).

[KM06]   Juha Kiviluoma and Peter Meibom. *Documentation of databases in the Wilmar Planning tool*. Risø-R-1554(EN). Roskilde, Denmark: Risø-National Laboratory, 2006. URL: http://www.wilmar.risoe.

dk / Deliverables / Wilmar % 20d6 _ 2 _ f _ DB _ doc . pdf (visited on 12/16/2009).

[Kon08]   Panos Konstantin. *Praxisbuch Energiewirtschaft: Energieumwandlung, -transport und -beschaffung im Liberalisierten Markt*. 2nd ed. VDI-Buch. Berlin: Springer-Verlag GmbH, Dec. 2008, p. 474.

[KW94]   Peter Kall and Stein W. Wallace. *Stochastic programming*. John Wiley, 1994.

[Lan08a]   Christoph Lang. "Quantifizierung von Marktmacht am deutschen Stromerzeugungsmarkt". In: *Marktmacht und Marktmachtmessung im deutschen Großhandelsmarkt für Strom*. 2008, pp. 67–90. DOI: 10.1007/978-3-8350-5524-7_6.

[Lan08b]   Christoph Lang. "Strukturelle Indikatoren für Marktmacht". In: *Marktmacht und Marktmachtmessung im deutschen Großhandelsmarkt für Strom*. 2008, pp. 53–66. DOI: 10.1007/978-3-8350-5524-7_5.

[Lem05]   Jacob Lemming. "Risk and investment management in liberalized electricity markets". PhD thesis. Department of Mathematical Modelling, Technical University of Denmark, 2005. URL: http://www2.imm.dtu.dk/pubdb/p.php?2823 (visited on 12/16/2009).

[Leu+08]   Florian Leuthold, Hannes Weigt, and Christian von Hirschhausen. *ELMOD – A Model of the European Electricity Market*. Electricity Markets Working Paper No. WP-EM-00. Chair for Energy Economics and Public Sector Management, Dresden University of Technology, July 2008. URL: http://papers.ssrn.com/sol3/papers.cfm?abstract_id=1169082 (visited on 12/16/2009).

[LF04]   Ying Li and Peter C. Flynn. "Deregulated power prices: comparison of volatility". In: *Energy Policy* 32.14 (Sept. 2004), pp. 1591–1601. DOI: 10.1016/S0301-4215(03)00130-7.

[Lij07]   Mark G. Lijesen. "The real-time price elasticity of electricity". In: *Energy Economics* 29.2 (Mar. 2007), pp. 249–258. DOI: 10.1016/j.eneco.2006.08.008.

[LS02]   Julio J. Lucia and Eduardo S. Schwartz. "Electricity Prices and Power Derivatives: Evidence from the Nordic Power Exchange". In: *Review of Derivatives Research* 5.1 (2002), pp. 5–50. DOI: 10.1023/A:1013846631785.

[Man07]   Manager Magazin. *Kunden verlassen Vattenfall*. July 2007. URL: http:
            //www.manager-magazin.de/unternehmen/artikel/0,2828,491939
            ,00.html (visited on 12/16/2009).

[Man+07]   Matteo Manera, Chiara Longo, Anil Markandya, and Elisa Scarpa.
            *Evaluating the Empirical Performance of Alternative Econometric Models
            for Oil Price Forecasting*. Fondazione Eni Enrico Mattei Working Pa-
            pers 69. Fondazione Eni Enrico Mattei, 2007. URL: http://ideas.
            repec.org/p/fem/femwpa/2007.4.html (visited on 12/16/2009).

[McK08]   McKinsey & Company. *Carbon Capture & Storage: Assessing the Eco-
            nomics*. Report. 2008.

[McL08]   James R. McLean. *TradeWind WP2.6 – Equivalent Wind Power Curves*.
            Project Report. July 2008. URL: http://www.trade-wind.eu/
            fileadmin/documents/publications/D2.4_Equivalent_Wind_
            Power_Curves_11914bt02c.pdf (visited on 12/16/2009).

[Mea+72]   Donella Meadows, Denis L. Meadows, Jorgen Randers, and William
            W. Behrens III. *The Limits to Growth – A Report for the Club of Rome's
            Project on the Predicament of Mankind*. New York: Universe Books,
            1972.

[Mei+06]   Peter Meibom, Helge V Larsen, Rüdiger Barth, Heike Brand,
            Christoph Weber, and Oliver Voll. *Wilmar Joint Market Model, Doc-
            umentation*. Risø-R-1552(EN). Risø National Laboratory, 2006. URL:
            http://www.wilmar.risoe.dk/Deliverables/Wilmar%20d6_2_b_
            JMM_doc.pdf (visited on 12/16/2009).

[Mez+07]   Jose L. Ceciliano Meza, Mehmet Bayram Yildirim, and Abu S.M. Ma-
            sud. "A Model for the Multiperiod Multiobjective Power Genera-
            tion Expansion Problem". In: *Power Systems, IEEE Transactions on* 22.2
            (2007), pp. 871–878. DOI: 10.1109/TPWRS.2007.895178.

[MG09]   Dominik Möst and Massimo Genoese. "Market power in the Ger-
            man wholesale electricity market". In: *The Journal of Energy Markets*
            2.2 (2009), pp. 47 –74.

[MM02]   Rémi Munos and Andrew W. Moore. "Variable Resolution Discretiza-
            tion in Optimal Control". In: *Machine Learning* 49.2 (Nov. 2002),
            pp. 291–323. DOI: 10.1023/A:1017992615625.

[Mon+04]   Christopher K. Monson, David Wingate, Kevin D. Seppi, and Todd S. Peterson. "Variable Resolution Discretization in the Joint Space". In: *Machine Learning and Applications, 2004. Proceedings. 2004 International Conference on*. 2004, pp. 449–455.

[Moo91]   Andrew W. Moore. *An introductory tutorial on kd-trees*. Technical Report No. 209, Computer Laboratory, University of Cambridge. Pittsburgh, PA, 1991.

[Moo92]   Douglas William Moore. "Simplicial Mesh Generation with Applications". PhD thesis. Cornell University, 1992.

[MS05]   Frederic H. Murphy and Yves Smeers. "Generation Capacity Expansion in Imperfectly Competitive Restructured Electricity Markets". In: *Operations Research* 53.4 (2005), pp. 646–661.

[MS86]   Robert McDonald and Daniel Siegel. "The Value of Waiting to Invest". In: *The Quarterly Journal of Economics* 101.4 (Nov. 1986), pp. 707–728.

[Müs04]   Felix Müsgens. *Market power in the German wholesale electricity market*. EWI Working Paper Nr 04.03. Energiewirtschaftliches Institut an der Universität zu Köln, May 2004.

[Müs06]   Felix Müsgens. "Quantifying Market Power in the German Wholesale Electricity Market Using a Dynamic Multi-Regional Dispatch Model". In: *Journal of Industrial Economics* 54 (Dec. 2006), pp. 471–498. DOI: 10.1111/j.1467-6451.2006.00297.x.

[Mye77]   Stewart C. Myers. "Determinants of corporate borrowing". In: *Journal of Financial Economics* 5.2 (Nov. 1977), pp. 147–175. DOI: 10.1016/0304-405X(77)90015-0.

[Nad64]   EA Nadaraya. "On Estimating Regression". In: *Theory of Probability and its Applications* 9 (Jan. 1964), pp. 141–142. DOI: 10.1137/1109020.

[New02]   David M. Newbery. "Problems of liberalising the electricity industry". In: *European Economic Review* 46.4-5 (May 2002), pp. 919–927. DOI: 10.1016/S0014-2921(01)00225-2.

[New08]   Nikki Newham. "Power System Investment Planning using Stochastic Dual Dynamic Programming". PhD thesis. Electrical and Computer Engineering, University of Canterbury, Apr. 2008. URL: http://hdl.handle.net/10092/1975 (visited on 12/16/2009).

[Nit07]   Joachim Nitsch. *Leitstudie 2007 – "Ausbaustrategie Erneuerbare En-ergien" Aktualisierung und Neubewertung bis zu den Jahren 2020 und 2030 mit Ausblick bis 2050.* Untersuchung im Auftrag des Bundes-ministerium für Umwelt, Naturschutz und Reaktorsicherheit (BMU). Feb. 2007.

[Nit08]   Joachim Nitsch. *Leitstudie 2008 – Weiterentwicklung der "Ausbaustrate-gie Erneuerbare Energien".* Untersuchung im Auftrag des Bundes-ministerium für Umwelt, Naturschutz und Reaktorsicherheit (BMU). Oct. 2008.

[Ock07]   Axel Ockenfels. "Marktmachtmessung im deutschen Strommarkt in Theorie und Praxis - Kritische Anmerkungen zur London Economics-Studie". In: *Energiewirtschaftliche Tagesfragen* 57.9 (2007), pp. 12–29.

[OI01]   Organisation for Economic Co-operation and Development (OECD) and International Energy Agency (IEA). *Competition in Electricity Markets.* OECD, 2001.

[OI05]   Organisation for Economic Co-operation and Development (OECD) and International Energy Agency (IEA). *Projected costs of generating electricity : 2005 update.* Paris: OECD/IEA, 2005.

[ÖI+08]   Öko-Institut, FZ-Jülich, DIW, and FhG-ISI. *Politikszenarien für den Klimaschutz IV - Szenarien bis 2030.* UBA-Forschungsbericht 001097. Dessau-Roßlau: Umweltbundesamt, 2008.

[Ols05]   Fernando Olsina. "Long-Term Dynamics of Liberalized Electricity Markets". PhD thesis. Instituto de Energía Eléctrica, Universidad Na-cional de San Juan, Sept. 2005.

[Ore05]   Shmuel S. Oren. "Ensuring generation adequacy in competitive elec-tricity markets". In: *Electricity deregulation: Choices and Challenges.* University of Chicago Press, 2005, pp. 388–414.

[OS03]   Arthur O'Sullivan and Steven M. Sheffrin. *Economics: Principles in Ac-tion.* Pearson Prentice Hall, 2003.

[Pad04]   Narayana P. Padhy. "Unit Commitment – A Bibliographical Survey". In: *Power Systems, IEEE Transactions on* 19.2 (2004), pp. 1196–1205. DOI: 10.1109/TPWRS.2003.821611.

[PB09]   Dimitri Perekhodtsev and Seth Blumsack. "International wholesale electricity markets and generators' incentives". In: *International Handbook On The Economics Of Energy*. Edward Elgar Publishing, Nov. 2009, pp. 624–649.

[Pil98]   Dragana Pilipovic. *Energy Risk*. McGraw-Hill, Jan. 1998.

[Pin01]   Robert S. Pindyck. "The Dynamics of Commodity Spot and Futures Markets: A Primer." In: *The Energy Journal* 22.3 (2001), pp. 1–29.

[Pin99]   Robert S. Pindyck. "The Long-Run Evolution of Energy Prices." In: *The Energy Journal* 20.2 (Apr. 1999), pp. 1–28.

[PM03]   Pierre-Olivier Pineau and Pauli Murto. "An Oligopolistic Investment Model of the Finnish Electricity Market". In: *Annals of Operations Research* 121.1 (July 2003), pp. 123–148. DOI: 10.1023/A:1023307319633.

[Poi09]   Point Carbon. *Carbon 2009 – Emissions trading coming home*. Report. Mar. 2009.

[Pow+02]   Warren B. Powell, Joel A. Shapiro, and Hugo P. Simão. "An Adaptive Dynamic Programming Algorithm for the Heterogeneous Resource Allocation Problem". In: *Transportation Science* 36.2 (2002), pp. 231–249. DOI: 10.1287/trsc.36.2.231.561.

[Pow+04]   Warren B. Powell, Andrzej Ruszczynski, and Huseyin Topaloglu. "Learning Algorithms for Separable Approximations of Discrete Stochastic Optimization Problems". In: *Mathematics of Operations Research* 29.4 (Nov. 2004), pp. 814–836. DOI: 10.1287/moor.1040.0107.

[Pow07]   Warren B. Powell. *Approximate Dynamic Programming: Solving the Curses of Dimensionality*. 1st ed. Wiley-Interscience, Sept. 2007.

[Pow09]   Warren B. Powell. "Merging AI and OR to Solve High-Dimensional Stochastic Optimization Problems Using Approximate Dynamic Programming". In: *INFORMS Journal on Computing* (Oct. 2009), to appear. DOI: 10.1287/ijoc.1090.0349.

[PP03]   Katerina P. Papadaki and Warren B. Powell. "An adaptive dynamic programming algorithm for a stochastic multiproduct batch dispatch problem". In: *Naval Research Logistics* 50.7 (2003), pp. 742–769. DOI: 10.1002/nav.10087.

[PS78]   Martin L. Puterman and Moon Chirl Shin. "Modified Policy Iteration
            Algorithms for Discounted Markov Decision Problems". In: *Manage-
            ment Science* 24.11 (July 1978), pp. 1127–1137.

[PT06]   Marc S. Paolella and Luca Taschini. *An Econometric Analysis of Emis-
            sion Trading Allowances.* Working Paper 341. National Centre of Com-
            petence in Research Financial Valuation and Risk Management, 2006.

[Rav+01]   Hans Ravn et al. *Balmorel: A Model for Analyses of the Electricity and
            CHP Markets in the Baltic Sea Region.* Balmorel main report. Mar. 2001,
            p. 70. URL: http://www.balmorel.com/Doc/B-MainReport0301.pdf
            (visited on 12/16/2009).

[RWE08]   RWE AG. *Do it. The new RWE.* Annual Report. 2008.

[SB98]   Richard S. Sutton and Andrew G. Barto. *Reinforcement Learning: An
            Introduction.* The MIT Press, Mar. 1998.

[Sch04]   Jochen Schröter. *Auswirkungen des europäischen Emissionshandelssys-
            tems auf den Kraftwerkseinsatz in Deutschland.* Diplomarbeit. Institut
            für Energietechnik, TU Berlin, Feb. 2004.

[Sch97]   Eduardo S. Schwartz. "The Stochastic Behavior of Commodity Prices:
            Implications for Valuation and Hedging". In: *Journal of Finance* 52.3
            (1997), pp. 923–73.

[SE04]   Alain Schmutz and Philipp Elkuch. "Electricity price forecasting: Ap-
            plication and experience in the European power markets". In: *Pro-
            ceedings of the 6th IAEE European Conference.* Zurich, Switzerland,
            2004.

[Sei+08]   Jan Seifert, Marliese Uhrig-Homburg, and Michael Wagner. "Dy-
            namic behavior of CO_2 spot prices". In: *Journal of Environmental Eco-
            nomics and Management* 56.2 (Sept. 2008), pp. 180–194. DOI: 10.1016/
            j.jeem.2008.03.003.

[Sen07]   Frank Sensfuß. "Assessment of the impact of renewable electricity
            generation on the German electricity sector - An agent-based simula-
            tion approach". Dissertation. Department of Economics and Business
            Engineering, Universität Karlsruhe (TH), 2007.

[Sen+07]   Frank Sensfuß, Mario Ragwitz, Massimo Genoese, and Dominik
            Möst. "Agent-based Simulation of Electricity Markets – A Literature
            Review". In: *Energy Studies Review* 15.2 (2007), pp. 19–47.

[SF94]   Gerald B. Sheble and George N. Fahd. "Unit commitment literature synopsis". In: *Power Systems, IEEE Transactions on* 9.1 (1994), pp. 128–135. DOI: 10.1109/59.317549.

[Sim+08]   Hugo P. Simão, Jeff Day, Abraham P. George, Ted Gifford, John Nienow, and Warren B. Powell. "An Approximate Dynamic Programming Algorithm for Large-Scale Fleet Management: A Case Application". In: *Transportation Science* 43.2 (2008), pp. 178–197. DOI: 10.1287/trsc.1080.0238.

[SM05]   Gorden Spangardt and Jürgen Meyer. "Risikomanagement im Emissionshandel". In: *Emissionshandel.* 2005, pp. 219–232. DOI: 10.1007/3-540-26685-2_12.

[Sol02]   Patricio del Sol. "Responses to electricity liberalization: the regional strategy of a Chilean generator". In: *Energy Policy* 30.5 (Apr. 2002), pp. 437–446. DOI: 10.1016/S0301-4215(01)00112-4.

[SP04]   Michael Z. Spivey and Warren B. Powell. "The Dynamic Assignment Problem". In: *Transportation Science* 38.4 (2004), pp. 399–419.

[Sta06]   Statkraft. *Stellungnahme von Statkraft zum Entwurf des NAP II.* 2006. URL: http://www.bmu.de/files/pdfs/allgemein/application/pdf/nap_stellungnahme_statkraft.pdf (visited on 12/02/2009).

[Sun+08]   Ninghong Sun, Ingo Ellersdorfer, and Derk J. Swider. "Model-based long-term electricity generation system planning under uncertainty". In: *2008 Third International Conference on Electric Utility Deregulation and Restructuring and Power Technologies.* Nanjing, China, 2008, pp. 1298–1304. DOI: 10.1109/DRPT.2008.4523607.

[SW07]   Derk J. Swider and Christoph Weber. "The costs of wind's intermittency in Germany: application of a stochastic electricity market model". In: *European Transactions on Electrical Power* 17.2 (2007), pp. 151–172. DOI: 10.1002/etep.125.

[Swi07]   Derk J. Swider. "Compressed Air Energy Storage in an Electricity System With Significant Wind Power Generation". In: *Energy Conversion, IEEE Transactions on* 22.1 (2007), pp. 95–102.

[Swi+07]   Derk J. Swider, Ingo Ellersdorfer, Matthias V. Hundt, and Al-
            fred Voss. *Anmerkungen zu empirischen Analysen der Preisbildung am
            deutschen Spotmarkt für Elektrizität.* Gutachten im Auftrag des Verban-
            des der Verbundunternehmen und Regionalen Energieversorger in
            Deutschland (VRE e. V.) Institute of Energy Economics and the Ra-
            tional Use of Energy, 2007.

[TB02]   Chung-Li Tseng and Graydon Barz. "Short-Term Generation Asset
            Valuation: A Real Options Approach". In: *Operations Research* 50.2
            (Mar. 2002), pp. 297–310. DOI: 10.1287/opre.50.2.297.429.

[TV01]   John N. Tsitsiklis and Benjamin Van Roy. "Regression Methods for
            Pricing Complex American-Style Options". In: *Neural Networks, IEEE
            Transactions on* 12.4 (2001), pp. 694–703. DOI: 10.1109/72.935083.

[UCT07]   Union for the Co-ordination of Transmission of Electricity(UCTE).
            System Disturbance on 4 November 2006. Final Report. 2007.

[Umw09]   Umweltbundesamt. *Datenbank "Kraftwerke in Deutschland" – Liste der
            sich in Betrieb befindlichen Kraftwerke bzw. Kraftwerksblöcke ab einer
            elektrischen Bruttoleistung von 100 Megawatt.* July 2009. URL: http:
            //www.umweltbundesamt.de/energie/archiv/kraftwerke_in_
            deutschland.pdf (visited on 12/16/2009).

[Van+97]   Benjamin Van Roy, Dimitri P. Bertsekas, Yuchun Lee, and John N.
            Tsitsiklis. "A Neuro-Dynamic Programming Approach to Retailer In-
            ventory Management". In: *Decision and Control, 1997., Proceedings of
            the 36th IEEE Conference on.* Vol. 4. 1997, pp. 4052–4057. DOI: 10.1109/
            CDC.1997.652501.

[Var06]   Hal Varian. *Intermediate Microeconomics: A Modern Approach.* 7th ed.
            New York: Norton, 2006.

[Vat08]   Vattenfall Europe AG. *The year 2008 in facts and figures.* Annual Re-
            port. 2008.

[Ven+02]   Mariano Ventosa, Rafael Denis, and Carlos Redondo. "Expansion
            Planning in Electricity Markets. Two Different Approaches". In:
            June 2002. URL: http://www.pscc-central.org/uploads/tx_
            ethpublications/s43p04.pdf (visited on 12/16/2009).

[Ven+05]   Mariano Ventosa, Álvaro Baíllo, Andrés Ramos, and Michel Rivier. "Electricity Market Modeling Trends." In: *Energy Policy* 33.7 (May 2005), pp. 897–913. DOI: 10.1016/j.enpol.2003.10.013.

[Vog04]   Klaus-Ole Vogstad. "A system dynamics analysis of the Nordic electricity market: The transition from fossil fuelled toward a renewable supply within a liberalised electricity market". Doctoral thesis. Faculty of Information Technology, Mathematics and Electrical Engineering, Norwegian University of Science and Technology (NTNU), Dec. 2004.

[Wat64]   Geoffrey S. Watson. "Smooth regression analysis". In: *Sankhya: The Indian Journal of Statistics, Series A* 26.4 (1964), 359–Ű372.

[Web05]   Christoph Weber. *Uncertainty in the Electric Power Industry*. Springer, 2005.

[Web+06]   Christoph Weber, Derk J. Swider, and Philip Vogel. "A Stochastic Model for the European Electricity Market and the Integration Costs fo Wind Power". In: *GreenNet-EU27 – Guiding a Least Cost Grid Integration of RES-Electricity in an Extended Europe* D5b (Feb. 2006), pp. 107–131.

[Wer00]   Rafal Weron. "Energy price risk management". In: *Physica A: Statistical Mechanics and its Applications* 285.1-2 (Sept. 2000), pp. 127–134. DOI: 10.1016/S0378-4371(00)00276-4.

[WH08]   Hannes Weigt and Christian von Hirschhausen. "Price formation and market power in the German wholesale electricity market in 2006". In: *Energy Policy* 36.11 (Nov. 2008), pp. 4227–4234. DOI: 10.1016/j.enpol.2008.07.020.

[Wie08]   Gerhard Wiedemann. *Handbuch des Kartellrechts*. 2nd ed. Beck Juristischer Verlag, Dec. 2008.

[WM08]   Rafal Weron and Adam Misiorek. "Forecasting spot electricity prices: A comparison of parametric and semiparametric time series models". In: *International Journal of Forecasting* 24.4 (Dec. 2008), pp. 744–763. DOI: 10.1016/j.ijforecast.2008.08.004.

[WP09]   WestLB and Potsdam-Institut für Klimafolgenforschung. *Deutsche Stromversorger – In der CO_2-Falle?* Studie. Sept. 2009.

[WS04]   Christoph Weber and Derk J. Swider. "Power plant investments un-
           der fuel and carbon price uncertainty". In: *Proceedings of the 6th IAEE
           European Conference*. Zurich, Switzerland, Sept. 2004.

[WV08]   Anke Weidlich and Daniel J. Veit. "A critical survey of agent-based
           wholesale electricity market models". In: *Energy Economics* 30.4 (July
           2008), pp. 1728–1759. DOI: 10.1016/j.eneco.2008.01.003.